LABORATORY MANUAL

Experiments in Fundamental Concepts of Biology

By Gideon E. Nelson with Gerald G. Robinson and Richard A. Boolootian:
  Fundamental Concepts of Biology Third Edition, 1974
By Gideon E. Nelson and Gerald G. Robinson:
  Study Guide for Fundamental Concepts of Biology Third Edition, 1974
Edited by Gideon E. Nelson and James D. Ray, Jr.:
  Contemporary Readings In Biology, 1974
Edited by Gideon E. Nelson and James D. Ray, Jr.:
  Biologic Readings for Today's Student, 1971
By Richard A. Boolootian:
  Introduction to Biological Concepts (set of twelve filmstrips), 1970
By Richard A. Boolootian:
  Human Biology Slide Sets: (six sets) Human Reproduction; Human Genetics; Human Development; Endocrine Function; Neurobiology; Control Mechanisms

LABORATORY MANUAL

Experiments in Fundamental Concepts of Biology

Third Edition

Gideon E. Nelson and Albert A. Latina
Biology Department
University of South Florida

John Wiley & Sons, Inc.
New York London Sydney Toronto

Copyright © 1968, 1970, 1974, by John Wiley & Sons, Inc.

All rights reserved. Published simultaneously in Canada.

Reproduction or translation of any part of this work beyond that permitted by Sections 107 or 108 of the 1976 United States Copyright Act without the permission of the copyright owner is unlawful. Requests for permission or further information should be addressed to the Permissions Department, John Wiley & Sons, Inc.

Library of Congress Cataloging in Publication Data:

Nelson, Gideon E
    Fundamental concepts of biology.

    Includes bibliographies.
    —Laboratory manual: experiments in Fundamental concepts of biology.
    1. Biology.　I. Robinson, Gerald G., joint author.　II. Boolootian, Richard A., joint author.
III. Latina, Albert A., 1936-　joint author.
IV. Title. [DNLM: 1. Biology–Laboratory manuals. QH317 N426L 1974]
QH308.2.N45　1974　　　574　　　73-9894
ISBN 0-471-63153-1
ISBN 0-471-63145-0 (suppl.)

Printed in the United States of America

10 9 8 7 6 5

PREFACE

This laboratory manual was prepared to accompany *Fundamental Concepts of Biology*, Third Edition by Nelson, Robinson, and Boolootian. However, the exercises are presented in such a way that they can be used with other general biology texts.

The objectives of the manual are as follows:

1. To facilitate an understanding of selected biological principles by the use of observation and experimentation.
2. To reinforce biological concepts presented in the textbook.
3. To stimulate further interest in Biology through the use of living organisms, biological "tools," and investigative procedures.

Each exercise contains questions that emphasize major points. Projects are listed at the end of most exercises for students who wish to do further investigations.

The manual has been revised to add descriptions and illustrations of several kinds of equipment used in modern research laboratories.

Gideon E. Nelson
Albert A. Latina

CONTENTS

EXERCISE 1   Characteristics of Life   1
EXERCISE 2   Life: Its Chemical Basis   17
EXERCISE 3   Life: Its Structural Basis   31
EXERCISE 4   Photosynthesis   45
EXERCISE 5   Respiration: Energy Harvest   53
EXERCISE 6   Transport of Materials   63
EXERCISE 7   Control Within Cells   75
EXERCISE 8   Control by Chemical Agents   83
EXERCISE 9   Control by Nervous Systems   89
EXERCISE 10   The Interaction of Control Systems: Homeostasis   99
EXERCISE 11   Communication and Behavior   105
EXERCISE 12   Asexual Reproduction   117
EXERCISE 13   Sexual Reproduction   127
EXERCISE 14   Development   135
EXERCISE 15   Structures Associated with Heredity   145
EXERCISE 16   Mendelian Genetics   153
EXERCISE 17   Human Hereditary Traits   159
EXERCISE 18   Population Genetics   165
EXERCISE 19   Adaptation and Variation   173
EXERCISE 20   Evolution: Study of Fossils   179
EXERCISE 21   Introduction to Ecosystems: A Local Plant Community   185
EXERCISE 22   An Analysis of a Terrestrial Community   189
EXERCISE 23   The Pond Community   195
                      Calendar for Preparation of Living Materials   207
                      Directions for Pithing a Frog   211

EXERCISE 1
Characteristics of Life

MATERIALS

Student Station

40 ml 5% copper sulfate solution
3 potassium ferrocyanide crystals (about 3 mm diameter)
5 ml dilute nitric acid (60 ml of concentrated acid to 1 liter distilled water)
5 ml dilute nitric acid (120 ml of concentrated acid to 1 liter distilled water)
1 g potassium dichromate crystals
2 Syracuse watch glasses
50-ml beaker
wooden splint
compound microscope
prepared slides of plant and animal tissues
dissecting needle or probe

Each Group of Four Students

slides and coverglasses
lens paper

dropping bottle of 4% methyl cellulose
mercury in dropper bottle

Class Stock

hay infusion (prepared 3 days in advance by adding dried weeds to a container filled with pond water)
containers for waste mercury

OBJECTIVES

1. To learn how to use microscopes properly.
2. To observe the osmotic plant and the artificial amoeba, models that demonstrate characteristics of life.

Part I
The Electron Microscope

Since the invention of the microscope in the seventeenth century, better instruments have continually been developed. From the simple single lens system of Leeuwenhoek to the sophisticated electron microscope, scientists have assembled these instruments to delve deeper into the mysteries of the microscopic world.

Electron microscopes allow biologists to view structures that are not visible with the conventional light microscopes (Fig. 1.1). Whereas the useful magnification of a light microscope is about 2500x, the resolving power of an electron microscope is many times that of the best light microscopes and thus allows greater useful magnifications. The resolving power refers to the ability of the microscope to make images distinct. If an object is not clear, increasing the magnification will not serve any useful purpose. Resolving power in optical microscopes is related to wavelengths of light. Thus the best resolving power with the optical microscope is limited by the shortest wavelength of light. Electrons have much shorter wavelengths than light waves. The electron microscope produces clear images at much greater magnifications than do light microscopes.

Basically the electron microscope differs from a light microscope in that the objects are not viewed directly (Fig. 1.2). Observations are made on a screen much like a television screen or by making photographs. Specimens must be subjected to a vacuum, because electron movement is affected by air molecules. They must be able to withstand this condition without altering their makeup drastically. The use of a vacuum precludes the possibility of viewing live specimens, a definite disadvantage of the electron microscope.

CHARACTERISTICS OF LIFE 3

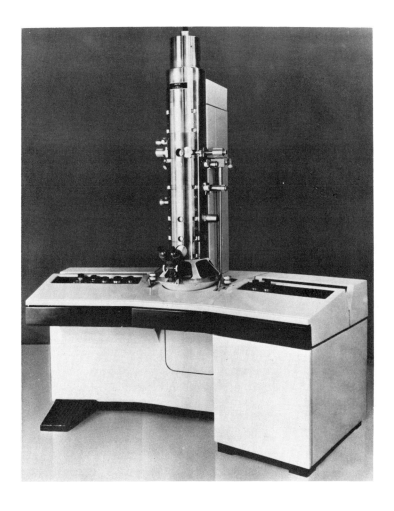

Figure 1.1   Electron Microscope (Courtesy Hitachi Perkin-Elmer).

The contrast necessary for distinguishing objects comes from the scattering of electrons. Most biological specimens scatter electrons poorly and require special preparations to overcome this problem. Thus tissues are commonly treated in solutions containing osmium atoms. Osmium impregnates the tissues and aids in scattering electrons. The osmium treated tissues are then embedded in a plastic or epoxy resin compound. The plastic block containing the specimen is sliced into ultra-thin sections so that electrons can pass through the specimen. Machines called microtomes are used for this purpose. Knives on microtomes are either made from glass or have a diamond cutting surface. The sections, 200 to 600 angstroms thick, are transferred to copper grids. These grids are placed in special holders in the electron microscope.

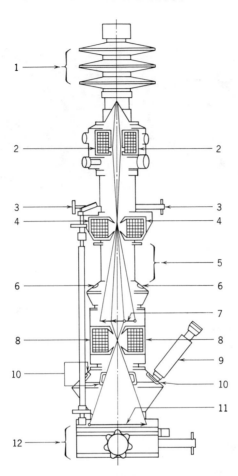

Figure 1.2 Schematic view of the electron microscope. (1) Electron gun. (2) Condenser lense. (3) Specimen chamber door. (4) Objective lense. (5) Objective barrel. (6) Viewing chamber for the intermediate image. (7) Intermediate image. (8) Projector lense (9) Ocular for the examination of the final image. (10) Viewing chamber for the final image. (11) Final image. (12) Photo assembly.

The textbook, *Fundamental Concepts of Biology*, and Exercise 2 of the laboratory manual show photographs of cellular contents made with the electron microscope. Compare the appearance of cells as seen by the electron miscroscope with cells as they are seen through the microscopes available in your laboratory.

It is not possible to make electron microscopes available to beginning students of biology. Most of them are very expensive, require elaborate physical facilities, and demand special training before they can be used properly.

The dissecting and compound microscopes are still very important instruments, and they are usually available in most laboratories. This exercise and others in the laboratory manual make use of these two kinds of microscopes. It is important that you pay special attention to learning to use them properly. Microscopes are useful only when operated correctly, and may be easily damaged if sufficient care is not taken.

Part 2
Use of the Compound Microscope

A microscope is a delicate and expensive piece of equipment. It is your responsibility to handle it with care, to use it correctly, and to store it properly when it is not in use. Treat a microscope as you would a fine watch or camera.

A. Proper Way to Hold a Microscope

There is a correct way to pick up and carry a microscope. Grasp the *arm* of the microscope (Fig. 1.3) with one hand and pick it up straight so that the long axis of the microscope is in a vertical position. Place the palm of your other hand directly under the microscope *base*. Hold the microscope in front of you in a comfortable position, taking care to be sure that it does not become tilted as you walk. If your microscope has a built-in illuminator, avoid touching the illuminator, since it will be hot after you have used the microscope for a while.

These details may seem unimportant, but they are necessary to prevent the *ocular* lens from falling out, and to prevent your accidentally hitting an obstacle with the microscope if it is held carelessly at your side.

B. Observation of a Permanent-Mount Slide

Low Power

Place the microscope on the table with the arm toward you. Be sure that the microscope *stage* is parallel to the surface of the desk and not slanted. The importance of this will become apparent later when you observe a temporary wet mount.

Locate the following: *ocular, coarse* and *fine adjustment knobs, tube, nosepiece* with *objectives, stage* with *clips, iris diaphragm* with or without a *condenser* (a revolving disk with various holes on some microscopes in place of an iris diaphragm), and *mirror* (one side concave, the other side flat) or *illuminator*.

Revolve the nosepiece until the low power objective (16-mm, 10X) is in line with the tube. A click will be heard when the objective is engaged properly. If this is not done correctly, you will see only your eyelashes against a black background. Some students never see beyond their eyelashes under the best conditions. You might as well give yourself a fighting chance! The total magnification of anything viewed through the microscope is obtained by multiplying the magnifying power of the objective in use by the magnifying power of the ocular. Hence with low power, which is the 10X objective, and a 10X ocular, the diameter of the object is magnified 100X.

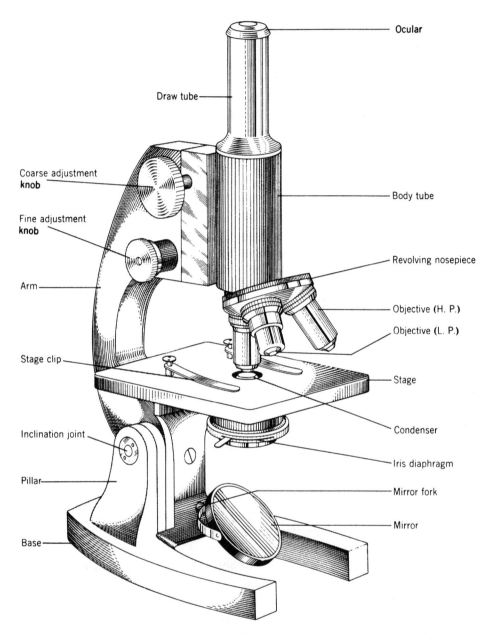

Figure 1.3 A compound microscope.

Using the concave side of your mirror, locate a light source. Move the mirror until a beam of light is reflected through the opening in the stage.

Select a permanent-mount slide from those provided and place it on the stage, making sure that the side on which the coverglass is mounted is up. Hold the slide in place with the stage clips, and position the slide so that the specimen to be observed is directly over the hole in the stage.

Look at the position of the objective lens relative to the coverglass on the slide. Turn the coarse adjustment knob until the objective lens is about 1/8 in. above the coverglass.

You are now ready to look through the microscope. *Keep both eyes open at all times.* Use whichever eye is more comfortable for you. Look through the microscope; turn the coarse adjustment knob toward you so that the objective lens is moved *away* from the coverglass. Do this until the object comes into focus. If you do not find the specimen, start over again. Never turn the coarse adjustment knob away from you while looking through the microscope. The objective may go through the coverglass and perhaps the slide as well, because some microscopes do not have an automatic stop. The crunching sound of an objective pushing through glass indicates careless, sloppy laboratory technique.

Once the object is in focus, it may be necessary to adjust the mirror or iris diaphragm to get the best possible light. A little practice will be required to accomplish this.

One hand should now be placed on the fine adjustment knob and left there for the entire viewing procedure. The other hand is used for moving the slide or drawing illustrations and writing comments. Move the fine adjustment knob a short distance toward and away from you until the object is focused clearly. You are now ready to make observations.

1. *Move the slide to the right. Which way does the object move, as viewed through the microscope? (It will be important to remember this.)*

Practice moving the slide until you can move it smoothly. Locate something in the field of vision that can be followed as the slide is shifted. Attempt to move the object completely around the edge of the field without losing it. (It will be particularly important to do manipulations of this type when you view live organisms.)

Rotate the fine adjustment knob as you move the slide.

2. *Why does the specimen not disappear immediately from view when you rotate the fine adjustment knob?*

When you are satisfied that you can operate the microscope successfully with the low power magnification, proceed to the high power (4 mm, 40X). When using high power, the object to be viewed must be at the center of the field because the high power objective magnifies only a small portion of what is seen with the low power.

High Power

Turn the nosepiece until the high power objective is in the proper position. Watch the objective closely from the side of the microscope as you turn the nosepiece. If the objective looks as though it will not clear the coverglass, move the fine adjustment knob to raise the lens slightly.

When the high power objective is in place, lower the objective until it almost touches the coverglass. Look through the ocular and focus upward using the fine adjustment knob. (The coarse adjustment knob must not be used with high power.) When the object is in focus, move the fine adjustment back and forth until the image is in sharp focus.

1. *What do you see with high power that you were not able to see with low power?*

2. *How much magnification do you get with the 40X objective and the 10X ocular?*

3. *Move the slide slowly. Attempt to follow a portion of the specimen around the edge of the field as you did with low power. What happens?*

You should never attempt to scan a slide with high power. Always do the preliminary observing with low power. Then proceed to high power to observe details.

CHARACTERISTICS OF LIFE 9

C. *Temporary Wet Mount*

Make a wet mount by the following method:

Start with a clean glass slide and coverglass. Place a drop of pond water on the center of the slide. With thumb and forefinger hold the coverglass and gently lower it so that an edge touches the drop of water (see Fig. 1.4). Lower the coverglass quickly to discourage the formation of air bubbles. Place the slide on the microscope and view with low power. Make use of the iris diaphragm to get the best light, not necessarily the brightest light. Less light should be used for fresh unstained material than for stained material.

1. *What do you observe?*

2. *Why must you continually regulate the fine adjustment to keep a moving organism in focus?*

3. *If the organism moves to the right as you view it through the microscope, which way must you move the slide to prevent losing it from the field?*

4. *What characteristics of life do you observe?*

10   EXPERIMENTS IN FUNDAMENTAL CONCEPTS OF BIOLOGY

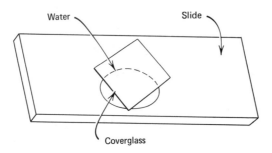

Figure 1.4

Add a drop of methyl cellulose to the edge of the coverglass and allow it to diffuse into the water on the slide. Focus on a microorganism that has been slowed by the methyl cellulose. Turn to high power, and record your observations.

Part 3
Use of the Stereoscopic Dissecting Microscope

Several exercises will suggest using a dissecting microscope. There are several reasons why this kind of microscope is very useful in magnifying small objects. The dissecting microscope is used as an important substitute for a magnifying glass, since it provides greater magnification of whole specimens, presents them in three dimensions, and allows space for manipulating and dissecting small specimens.

A. *Proper Way To Hold a Dissecting Microscope*

Follow a procedure for handling the dissecting microscope that is similar to the one you followed for the compound microscope. Notice the presence of two oculars which fall out if you are not careful. Also locate the white plate on the base, which is held loosely in place.

B. *Observations of Whole Specimens*

Place the microscope on a table with arm toward you. Locate the following (Fig. 1.5): oculars, ocular focusing sleeve, body clip, ocular tubes, focusing knob, stage, plate on stage.

Your microscope will either have a power control knob for changing magnifications or a revolving or sliding nose piece with two or three sites of objectives mounted on it. Find the lowest magnification (lowest number on power control knob), and begin your observations with this magnification.

Turn on your light source. This may be a gooseneck lamp or a special lamp designed for use with a dissecting microscope. Direct the light so that it is concentrated on the white plate on the base. If your microscope has a clear glass plate, place a piece of white paper on it.

Choose any one of the specimens provided (preferably one that is not flat), and place it on the plate. Look through the ocular lenses and rotate the focusing knobs until the specimen is seen. If this is your first experience in using a three-dimensional viewer, it will probably take a while to adjust your eyes. If things don't seem quite right (perhaps you are seeing a double image), try the following. The oculars are mounted in a slide so that they may be moved. Slide them back and forth as you look through the microscope until the distance apart seems comfortable for your eyes. Next check the oculars to see if one can be focused independently of the other by rotating a focusing sleeve. If the focusing ocular is on your left, close your left eye and look through the microscope with your right eye. Focus on a part of the specimen. Close your right eye and open your left eye. Rotate the focusing sleeve until the same part of the specimen is in sharp focus. Open your right eye and hope for the best.

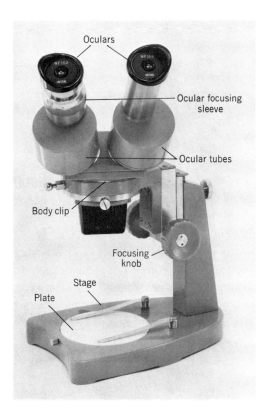

Figure 1.5   Stereoscopic microscope (Courtesy Unitron Instrument Company).

Notice the depth of field that is in focus. Rotate the focusing knobs back and forth and observe the change in what can be seen. With the aid of a dissecting needle or forceps, attempt to manipulate your specimen. Move it about without losing it from the field of view. Gain practice in directing the forceps or needle to predetermined places on the specimen. This kind of technique will be important for later exercises. Rotate the power control knob or change the objectives to a higher magnification.

1. What happens to the field of vision as you increase the magnification?

2. What change do you notice about the depth of the field in focus as the magnification is increased?

Part 4
Models of Life: Osmotic Plant and Artificial Amoeba

What is life? A more difficult question would be hard to imagine. Biologists make an attempt to define life as it is defined in the first chapter of *Fundamental Concepts of Biology* (by Nelson, Robinson, and Boolootian), but they leave more exact definitions to Webster and the philosophers. Practically everyone agrees that an oak tree growing in the field and a rabbit running through the brush are alive. However, distinctions between living and nonliving things are not always that clear. For example, biologists still debate whether or not viruses are living organisms.

There is another question sure to cause debate among some biologists: How can one differentiate between a plant and an animal? This seems like a ridiculous question until you stop to consider that there are many organisms with both plant and animal characteristics. There have been serious proposals to create a third kingdom of living things to accommodate organisms that are not clearly plant or animal.

What is life? What is a plant? An animal? These are only a few of the countless questions that arouse the curiosity of biologists. The biologist, in seeking answers to his questions, makes use of the scientific method of inquiry. *This requires careful observations, searching, precise asking of questions, formulating testable hypotheses (educated guesses), and, finally, carefully controlled experimentation.* The hypotheses are either accepted or rejected depending on the results of the experiment. As you work the exercises in this manual, you should apply this method of inquiry continually and systematically.

Recall how difficult it is to define life. Even the characteristics we assign to living organisms are not themselves infallible. It is possible to demonstrate inorganic systems that exhibit characteristics similar to living systems. What then is the difference? That is the question for you to consider as you observe the following experiments. Write down as much as you can about what you observe.

A. Osmotic Plant

Place 3 crystals of potassium ferrocyanide in a 50-ml beaker containing 40 ml of 5% copper sulfate solution. Watch closely for 5 minutes.

1. *What do you surmise must have happened to cause the yellow crystals to change color?*

2. *What characteristics similar to living organisms do you observe?*

The growth that you see is produced by a process called osmosis. This phenomenon will be studied later in another exercise. For now we shall say that an inorganic membrane is formed and water moves through it into the rapidly dissolving crystal. This movement of water causes the membrane to swell until it ruptures at its weakest point. As soon as this happens, the potassium ferrocyanide solution rushes out; it then reacts with copper sulfate to form a new membrane.

3. *Suppose you say that since this "plant" exhibits growth, a characteristic of life, it is alive. How might you test this hypothesis? (Hint: Compare this type of growth with growth in plants and animals.)*

4. *Does the osmotic plant respond to a stimulus? If so, does it respond in the same way that plants and animals respond?*

14   EXPERIMENTS IN FUNDAMENTAL CONCEPTS OF BIOLOGY

5. Mount a small amount of the osmotic plant on a slide in a drop of copper sulfate solution, and observe through the microscope. Describe what you see.

6. Compare the microscopic appearance of the osmotic plant with a prepared slide of a plant tissue and one of an animal tissue. Carefully describe in your own words what you see on each slide.

7. How do plant and animal tissues differ?

8. In what ways are plant and animal tissues similar?

9. Do there appear to be more differences or similarities?

10. How do they compare with the miscroscopic appearance of the osmotic plant?

11. *Based on your observations, would you accept or reject the hypothesis that the osmotic plant is alive?*

B. Artificial Amoeba

Two different concentrations of nitric acid have been provided. Add 5 ml of one concentration to one Syracuse watch glass and the other concentration to another Syracuse watch glass. With a medicine dropper introduce a drop of mercury about 1/4 in. to each of the watch glasses. Caution! Do not spill any mercury because it is extremely toxic. Notify your instructor if any is spilled. Use a wooden splint to add a few crystals of potassium dichromate to each watch glass. The crystals should be added a short distance away from the drop of mercury. Watch closely as the crystals dissolve. As soon as the diffusing chemical reaches the mercury, the action begins. Watch for a while, and then describe what you see.

The action occurs because the potassium dichromate in nitric acid lowers the normally high surface tension of the mercury, causing it to flow.

What characteristics of life do you observe in this nonliving system?

When you finish this experiment, pour the mercury into the container provided for this purpose. *Do not pour mercury into the sink.* It will ruin the plumbing!

SUMMARY

1. The microscope, when used properly, becomes an effective tool for the scientist to learn more about living organisms.
2. The experiments involving the osmotic plant and the artificial amoeba point out that at certain levels it is difficult to say what is alive and what is not alive.
3. By comparing plant and animal tissues, it becomes evident that there are basic similarities between plants and animals.

REFERENCES

Wald, George, Peter Albersheim, John Dowling, Johns Hopkins III, and Sanford Lack, *Twenty-Six Afternoons of Biology,* Addison-Wesley, Reading, Mass., 1962.

Weisz, Paul B., *Laboratory Manual in the Science of Biology,* Second Edition, McGraw-Hill, New York, 1963.

EXERCISE 2

MATERIALS

Student Station

8 test tubes in rack
test tube brush
100-ml graduated cylinder
2 100-ml beakers
microscope

Each Group of Four Students

50-ml bottle of concentrated glucose solution
10% salt solution in dropper bottle
starch solution in dropper bottle
iodine solution (0.3 iodine crystals, 1.5 g potassium iodide, 100 ml water) in dropper bottle
Millon reagent in dropper bottle
50 ml Benedict's solution
5 g Sudan IV

50 ml diphenylamine reagent (Mix 97.25 ml glacial acetic acid with 2.75 ml concentrated H_2SO_4. Add 1 g of fresh diphenylamine and stir until dissolved. Add 100 ml water.)

Irish potato

apple

10 g of chopped liver

1 egg, separated

20 ml salad oil

20 ml whole milk

20 finely ground fresh or dry roasted peanuts in 50 ml of water

20 ml blood cells in isotonic saline

1 g yeast in 20 ml of water

50 ml 5% glycogen solution

5 g copper sulfate crystals

3/4–in. styrofoam balls in four colors: black, yellow, green, white

pipe cleaners

water bath

cork borer

microscope slides and coverglasses

3–in. piece of dialysis tubing

25 ml alkaline glycerol (add 10 drops of conc. sodium hydroxide to glycerol)

Phenolphthalein indicator solution in dropper bottle

string

Class Stock

living *Elodea*

paraffin or beeswax

3/4–in. one-hole rubber stopper

12–in. piece of glass tubing

ring stand and clamp

carrot

OBJECTIVES

1. To illustrate how atoms are linked to form molecules and compounds.
2. To demonstrate the presence of organic compounds in various biological materials.
3. To observe the functioning of diffusion and osmosis.

Set up experiments for Part 3 at this time and return to them near the end of the laboratory period.

**Part 1
Molecular Models**

For this exercise assume that the colored styrofoam balls represent atoms that are color-coded as follows:

>   black   = carbon
>   yellow = nitrogen
>   green = hydrogen
>   white = oxygen

Pipe cleaners represent bonds that link atoms. Thus a model of oxygen consists of two white styrofoam balls joined by two pipe cleaners.

By means of this technique, construct the following molecules: (*a*) water; (*b*) carbon dioxide; (*c*) ammonia; (*d*) atmospheric nitrogen; (*e*) a carboxyl group; (*f*) an amino group; (*g*) glycine amino acid; (*h*) glucose. Refer to Chapter II in *Fundamental Concepts of Biology* for these formulas.

1. *In which of the models you constructed do pipe cleaners stand for covalent bonds?*

2. *Which of the models represent both a molecule and a compound?*

3. *Which of these substances contains the largest number of atoms?*

20 EXPERIMENTS IN FUNDAMENTAL CONCEPTS OF BIOLOGY

4. *Which of the atoms used here contains the largest number of reactive bonds?*

5. *Which of the models contains atoms linked by double bonds?*

Part 2
Organic Compounds in Biological Materials

A. *Carbohydrates*

1. Test for simple sugars. Pour 5 ml of Benedict's solution into a test tube, add 15 drops of glucose solution, and place the tube in a boiling water bath for several minutes. The reddish precipitate confirms the presence of sugar in the solution. Benedict's solution contains copper sulfate, which is reduced to cuprous oxide by certain chemical groups in the sugar. The color varies from greenish (small amount of sugar) to brick red (large amount of sugar).

Test the following substances for the presence of sugar and record your results:

 a. Chopped liver
 b. Egg white
 c. Milk
 d. Apple scrapings

2. Test for starch. Add a few drops of iodine solution to a test tube containing 5 ml of tap water. Note the typical iodine color. Add a few drops of starch solution and observe the color change.

Test the following substances for the presence of starch and record your results:

 a. Potato scrapings in water
 b. Apple scrapings in water
 c. Glycogen (animal "starch")
 d. Egg white

B. Lipids

Sudan IV, a dye, dissolves in fatty materials and turns a reddish-orange color. Verify this test in the following way: Add a few Sudan IV grains to some water in a test tube. Shake gently. Add a small amount of salad oil, mix again, and observe the results.

Test the following substances in a similar way for the presence of lipids and record your results. Let each test stand for *5 min* before reading.

    a. Whole milk
    b. Egg yolk
    c. Egg white
    d. Finely ground peanuts

C. Proteins

Millon's reagent produces a reddish color in the presence of tyrosine, an amino acid. Most proteins contain this amino acid, hence the Millon test is used as an indicator for the presence of proteins.

Test this reaction by adding a few drops of Millon's reagent to 5 ml egg white in a test tube. Place the tube in a boiling water bath for a few minutes and note the color reaction.

Perform this test on the following materials and note your results:

    a. Whole milk
    b. Finely ground peanuts
    c. Potato scrapings

1. *If you had a test for* polypeptides *which of the above materials would give a positive reaction?*

2. *Based on the results obtained in the preceding tests, what might you conclude about the presence of proteins in plant and animal cells?*

3. *When you tested egg white for protein at the beginning of this exercise, what physical change took place in the egg white while it was in the water bath? Discuss the possibility of using this test as an additional test for protein.*

D. *Nucleic Acids*

The diphenylamine test indicates the presence of DNA by producing a blue color and indicates the presence of RNA by a green color.

Measure 3 ml water into one test tube, 3 ml yeast suspension in another, and 3 ml of blood solution into a third. Add 5 ml of diphenylamine reagent to each tube and place them in a near boiling water bath for 15 min. Allow the tubes to cool and observe the results.

1. *Why was the test performed on a tube of plain water?*

2. *Could you conclude whether yeast contains more RNA than DNA?*

3. *What did the reaction indicate about blood cells?*

4. *Would you expect a positive test with all biological substances? With all cellular materials? Explain.*

LIFE: ITS CHEMICAL BASIS   23

Part 3
Separation of Biological Materials

To study the compounds making up cells and tissues, it is often necessary to separate them from each other. A variety of techniques and equipment have been developed for isolating and purifying substances in mixtures of this kind. The techniques include various types of *chromatography* and *electrophoresis*.

Chromatography is a separation technique that may be performed in many different ways. Generally the mixture to be analyzed is first dissolved in an appropriate solvent such as alcohol. It is then placed in contact with a porous material such as a strip of filter paper or is poured through a tube packed with a finely powdered absorbing substance. In either case, as the mixture passes over the porous material, the components in it move at different rates of speed and thus gradually separate from one another. These techniques are especially useful for separating amino acids in a mixture, simple sugar and sugar derivatives, mixtures of lipids, and chlorophyll pigments.

The technique of *paper chromatography* forms part of Exercise 4 where it will be used for separating chlorophylls *a* and *b*.

A chromatographic technique widely used for separating amino acids or proteins is that of ion exchange columns. These are glass tubes packed with resins that react with electrical charges on other molecules. This is especially useful for separating proteins, since these large molecules often have distinctive electrical properties. As

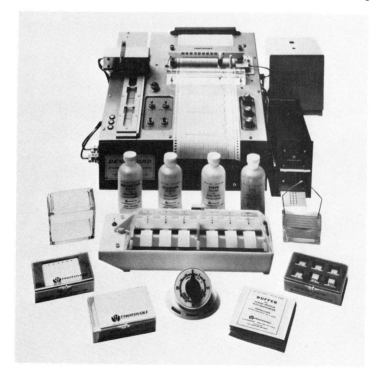

Figure 2.1   Electrophoresis apparatus (Courtesy Photovolt)

the mixture passes through the column, the components become separated because of their different electrical charges. Consequently, each one drips from the bottom of the column at a different time and can be collected in a separate container. See

Electrophoresis is another widely used technique. An electrical field is set up in a conducting medium, with the result that charged particles, such as proteins, migrate toward the electrode having an opposite charge. Since the net charge on different proteins vary, some migrate more rapidly than others and are thus separated. The conducting medium may be a variety of materials including filter paper, starch gel, and agar. See Figure 2.1. Manufacturers sell a variety of electrophoresis equipment, some of it capable of recording bands of proteins and their relative concentrations as they migrate in the electrical field.

Part 4
Movement of Molecules: Diffusion and Osmosis

A. Diffusion

Fill a small test tube with tap water. Drop a piece of copper sulfate into the water. Set aside and observe at intervals during the remainder of the lab period.

1. *What was the appearance of the solution when the copper sulfate was dropped into the water?*

2. *Describe the appearance of the solution an hour or so later. Explain.*

3. *Which diffuses more rapidly within a cell, molecules of glucose or molecules of a protein? Why?*

4. *Could diffusion occur in other states of matter, such as a gas into a gas, liquid into a liquid, or gas into a liquid?*

B. *Osmosis*

1. *Elodea* leaf.

Place a small *Elodea* leaf on a microscope slide under a coverglass. Observe under low power and move the slide until you find a portion of the leaf with one or two layers of cells. Focus on the area under medium power until you have a good view of cells showing their chloroplasts and cytoplasm. Gently place a few drops of salt solution beside the edge of the coverglass. Within several minutes a change should occur within some of the cells. After this occurs, answer the following questions:

1. *What happened to the contents of the cells?*

2. *Explain your answer to number 1.*

3. *Can you reverse the reaction in the* Elodea *leaf by rinsing it in fresh water? (Try it and see!)*

2. Carrot (Demonstration)

Remove a core from the center of a carrot with the cork borer and set up the apparatus as shown in Fig. 2.2. Pour glucose solution into the cavity in the carrot. Be careful not to spill any on the outside of it. Tightly seal the one-hole stopper in the carrot. Submerse the lower half of the carrot in water. At 30-min intervals observe the level of water in the glass tubing, and test the water in the beaker with Benedict's solution. Record your observations.

26 EXPERIMENTS IN FUNDAMENTAL CONCEPTS OF BIOLOGY

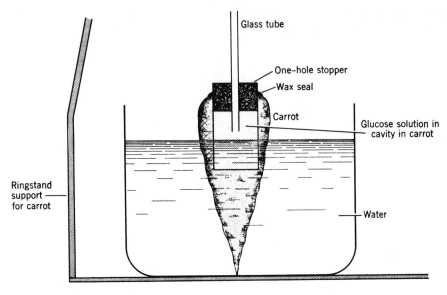

Figure 2.2 Carrot osmometer apparatus.

Observations

0 min

30 min

60 min

90 min

120 min

1. *Why did water move into the carrot?*

2. *Why did sugar move out of the carrot?*

3. Which diffused more rapidly, sugar or water molecules?

4. Is this phenomenon related to how carrots might obtain water from the soil? Explain.

3. Osmosis through an artificial membrane.

Rub dialysis membrane between your fingers in water until it becomes a tube. With a string tie one end of the tubing as tightly as possible. Partially fill the tubing with alkaline glycerol and securely tie the end. Rinse the tubing under running water. Place the tubing with the colorless alkaline glycerol into a beaker of water containing 5 drops of phenolphthalein. Phenolphthalein is pink in an alkaline pH.

1. Describe the appearance of the water in the beaker after the tubing was placed in it.

2. What caused the color change?

3. Was the color change a result of osmosis or diffusion?

4. *Describe the condition of the bag after 50 min and explain the change.*

5. *How does the function of this membrane differ from a living membrane?*

4. Application of the dialysis principle.

A dialysis membrane is properly termed "semipermeable," since the membrane itself does not actively control what passes through it. Molecules smaller than the tiny pores in the membrane may diffuse through it, but larger molecules cannot. An apparatus called a dialyzer is frequently used for concentrating colloidal (large particle) solutions and for adjusting ionic concentrations in solutions.

In recent years the principle of dialysis has been applied in constructing the artificial kidney machine. See Figure 2.3. In this apparatus, blood flows from the patient into coils of dialysis tubing immersed in a washing solution. Waste products diffuse from the blood into the solution; also, substances may be added to the blood via the solution if desired. Since dialysis is a relatively slow process, patients often must remain connected to the machine for many hours.

SUMMARY

The exercises you have performed illustrate some of the basic chemical and physical features of living matter. Molecules and compounds are built of repetitive atomic units. Compounds in living matter may be exceedingly large, complex, and uniquely constructed. Thus it is possible to detect their presence by using an appropriate chemical test. As you have observed, carbohydrates, lipids, proteins, and nucleic acids are found in a variety of biological materials. They may be separated for study and analysis by a variety of laboratory techniques including chromatography and electrophoresis.

Diffusion and osmosis are important functions in living matter in relation to the passage of materials through cell membranes.

REFERENCES

Korn, Robert W., and Ellen J. Korn, *Investigations into Biology,* John Wiley and Sons, New York, 1965.
Lawson, Chester A., and Richard E. Paulson, *Laboratory and Field Studies in Biology,* Holt, Rinehart and Winston, New York, 1960.

Humphrey, Donald G., Henry Van Dyke, and David L. Willis, *Life in the Laboratory,* Harcourt, Brace and World, New York, 1965.

PROJECTS

1. Repeat the carrot experiment with different concentrations of glucose and with other materials such as salt solutions.
2. Run a series of carrots with the same concentration of glucose solution to determine if carrots differ with respect to osmotic role.
3. Construct an osmotic cell with a 3-in. length of dialysis tubing. This can be set up similarly to the carrot experiment. Various solutions can be tested for their diffusion rates.

Figure 2.3 Artificial kidney machine (Courtesy National Kidney Foundation).

EXERCISE 3
Life: Its Structural Basis

MATERIALS

Student Station

Syracuse watch glasses
pith
razor blades
compound microscope
dissecting tools (scissors, scalpel, probe, needle)
dropper

Each Group of Four Students

Elodea plant
bean plant
string
slides and coverglasses
1% aqueous safranin solution
toothpicks
methylene blue in dropper bottle
water in dropper bottle

prepared slides of mammalian tissues:
    blood smear
    smooth muscle
prepared slides of various organs

OBJECTIVES

1. To observe some basic structures found in plant and animal cells.
2. To study form-function relationships in cells, tissues, and organs.

Part 1
Preparation of Permanent Slides

    Some of the slides of specimens that you will use in various exercises in this laboratory manual will be ones that you will make. These will be temporary slides, since the material on them will be discarded at the end of the laboratory period. Temporary slides are made quickly, and they do not require elaborate staining procedures. Their usefulness lies in the rapidity with which they can be made. The disadvantages of the temporary mount include the short time they can be preserved and the failure of optimally staining certain cellular structures.

    Most of the slides that you will use will be made commercially, and they may be used repeatedly. The permanent type slide may be made in the general biology laboratory, but the procedures are time consuming; and they require experience in handling the equipment properly. In the description of permanent slide production that follows, pay particular attention to the number of steps and the time involved. Keep these points in mind the next time you place a slide on your microscope.

    The first step in preparing permanent slides is to kill and fix the tissue that is to be examined. The cells in the tissue must be killed quickly to prevent deterioration, and then they must be fixed to preserve the cellular characteristics as closely as possible to the living condition. Solutions of alcohol, chloroform, glacial acetic acid, and picric acid are commonly used to kill and fix tissues.

    After approximately 24 hrs the fixative is washed out. The next step is to remove water from the tissues. This dehydration process involves a series of increasing concentrations of alcohol.

    When all the water is removed, the tissue is embedded in paraffin heated to a liquid. The paraffin infiltrates the tissues. This process requires many hours to insure complete infiltration. When ready, the paraffin containing the tissue is poured into a shallow container called a boat. The paraffin is then hardened by cooling it rapidly in ice water. The paraffin block containing the tissue is now ready to be sectioned on a rotary microtome.

LIFE: ITS STRUCTURAL BASIS   33

Figure 3.1   Rotary microtome (Courtesy Ivan Sorvall Inc.).

A microtome (see Fig. 3.1) is a precision slicing machine that cuts tissues into extremely thin sections. Thin sections are important for proper staining and for allowing light to penetrate so that cells can be observed with the microscope. The paraffin block is mounted on the microtome, and it is advanced forward 10 to 15 microns (1 micron = 1/1000 mm) with each turn of the handle. The sections stick to one another as they are cut, producing a ribbon that looks like a tapeworm. A few of these sections are placed on a glass microscope slide. The slide is heated causing the paraffin and the tissue to stick to it.

The paraffin is now removed by dipping the slide in a solvent (e. g., xylol). The next step is to stain the tissues. Frequently used stains are safranin, eosin, fast green, and hematoxylin. The last step is to place a drop of a sticky substance such as balsam or piccolyte on the tissue. A cover glass seals the tissue and makes the slide permanent.

Part 2
The Ultracentrifuge

Later in this exercise you will look at the structural characteristic of cells using microscope slides. Much has been learned in recent years about the various parts of cells by using a variety of different techniques. One technique (which deals with the use of prepared microscope slides of cells) has already been described in Part I. Another technique uses the ultracentrifuge to separate cellular components. The cells must first be broken open by various techniques to release their contents. The contents are placed into a special solution such as sucrose to keep the cellular components from being altered.

Basically all centrifuges operate in the same way, from the simplest one found in general biology laboratories to the sophisticated ultracentrifuges of the research laboratories.

The materials to be centrifuged are placed in special holders, and then they are spun at high speed to separate the various components. Various fractions have different densities, and therefore separate at different rates. Ultracentrifuges which look like modern-day washing machines (see Fig. 3.2) operate at speeds greater than 20,000 rpm. The extremely high speed allows the separation of substances that are not possible with lower speed centrifuges. Ultracentrifuges have been of immeasurable benefit to the research scientist. The ultracentrifugation of biologically active substances has yielded the separation of things such as purified proteins. In fact, some proteins were not even known to exist until the advent of the ultracentrifuge.

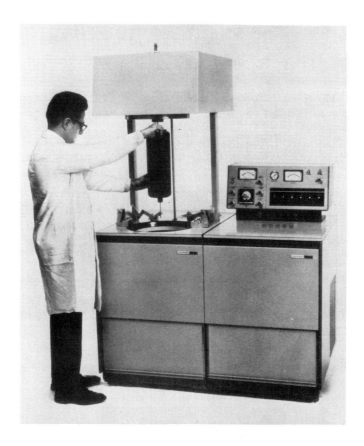

Figure 3.2 Ultracentrifuge (Courtesy Electro-Nucleonics, Separation & Analytical Systems Division).

Part 3
Plant Cells and Tissues

A. Elodea *Leaf Whole Mount*

Mount a small leaf from the tip of an Elodea plant in a drop of water. Place a coverglass over it. It is important that the leaf does not become dry. Add water to the edge of the coverglass as needed. Examine the slide with the low power of the microscope. Locate the cell wall. All living cells have a plasma membrane. However, because it is very thin and close to the cell wall, it cannot be seen.

1. *Are all the cells of the same size and form?*

2. *How many layers of cells are there?*

3. *Are there many small vacuoles or one large one in each cell?*

4. *What is the function of vacuoles? (See* Fundamental Concepts of Biology, *Chapter 3.)*

5. *Can you see any connection between the size and shape of the vacuoles in the cells and their function?*

6. Where are the chloroplasts located?

7. Examine the cells closely until you find a cell in which the chloroplasts are moving. Observe under high power. Can you account for the movement?

8. While using low power, try to locate the nucleus. Why is it difficult to find?

B. Leaf Cross Section

Study a permanent-mount slide of a leaf cross section under the microscope, or prepare your own slide by the following method:

1. Cut a 2-in. length of pith in half. See Fig. 3.1.
2. Place a small portion of a leaf such as a bean leaf between the two pieces of pith, as illustrated in Fig. 3.4. Tie the two halves securely with a piece of string.
3. Prepare thin cross sections by cutting the material with a straight razor or a single-edge razor blade. See Fig. 3.5.
4. Float the sections on water until ready for use.
5. Prepare a temporary wet mount according to instructions given in Exercise 1. (If stained sections are desired, sections may be placed in a 1% aqueous solution of safranin for 2 min and then rinsed in water before mounting on a slide.)

Observe the slide with the low power of the microscope. Locate the nucleus, chloroplasts, cell wall, and vacuole if present.

1. What is the shape of the cells?

LIFE: ITS STRUCTURAL BASIS 37

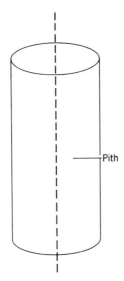

Figure 3 3

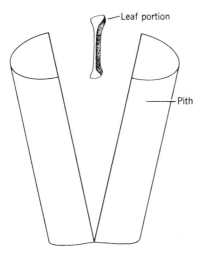

Figure 3.4

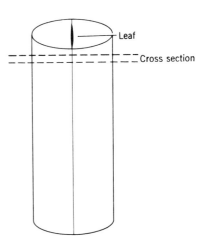

Figure 3.5

2. Recall that cells are 3-dimensional. How can you show that the cells you see have depth?

3. Are there spaces between the cells?

4. How do the cells fit together?

5. Are all the cells alike?

6. Describe what you see.

7. Are there groups of similar cells?

8. What do you call a group of similar cells specialized for the performance of a common function?

9. *With the help of a textbook, locate epidermis, stomata, mesophyll consisting of spongy and palisade layers, and vascular bundles. Are the stomata located on the surface of the epidermis or are they sunken?*

10. *Can you speculate about the environment in which this kind of plant grows by the position of the stomata?*

11. *Note the position of the vascular bundles. The bundles are primarily made up of xylem and phloem. Xylem are water-conducting cells, and phloem functions in moving food. How do food and water get to and from every cell even though xylem and phloem do not come in contact with every leaf cell?*

Part 4
Animal Cells and Tissues

A. *Cheek Epithelium*

Gently scrape the inside of your cheek with the flat end of a toothpick. There need not be any visible mass of tissue on the toothpick. Smear the scrapings on a slide. Add a drop of methylene blue and a coverglass. Examine with the low power of the microscope.

1. *What is the shape of the cells?*

2. *Are the cells thick or flattened?*

3. *Examine closely a cluster of cells. What geometric figure do the individual cells in the cluster approximate?*

4. *Can you offer any possible significance for this shape?*

5. *What is the function of epithelia?*

6. *Would you say that the form of the cells you see give any indication of the function of these cells? Explain.*

7. *What cell structures do you see?*

8. *What are some of the reasons that would account for the inability to see all of the cellular structures you would expect to find in an animal cell? (Refer to* Fundamental Concepts of Biology, *Chapter 3 on cell structures.)*

9. *What are some differences between plant and animal cells?*

LIFE: ITS STRUCTURAL BASIS   41

B. *Blood Smear*

Observe a blood smear slide with the low and high powers of the microscope.

1. *Are all the cells alike?*

2. *Can you distinguish between red and white blood cells?*

3. *Which are more numerous?*

4. *Why is blood considered to be a tissue?*

C. *Muscle*

Examine a prepared slide of smooth muscle.

1. *How are the cells arranged in muscle tissue?*

2. *The cells are spindle shaped. Of what signifiance is this insofar as the function is concerned?*

Part 5
Organs

Organs are aggregations of tissues in some particular structural form which perform a specific function. As time permits, study prepared sections of various organs such as liver, lungs, and stomach.

1. *Can you distinguish groups of similar cells in the organ?*

2. *What is the primary function of the organ?*

3. *How does the arrangement of the tissues allow for successful accomplishment of the primary function of the organ?*

SUMMARY

In this exercise you learned that cells have certain basic structural similarities; however, there is considerable diversity in cell forms. The diversity did not come about accidentally, since it is evident that the form is closely correlated with the function of the cell. The same form-function relationship holds true at higher levels of organization such as in tissues and organs.

REFERENCES

Humphrey, Donald G., Henry VanDyke, and David Willis, *Life in the Laboratory,* Harcourt, Brace and World, New York, 1965.

Lawson, Chester A., Ralph W. Lewis, Mary Alice Burmester, and Garrett Hardin, *Laboratory Studies in Biology,* W. H. Freeman and Co., San Francisco, 1955.

Morholt, Evelyn, Paul F. Brandwein, and Joseph Alexander, *A Sourcebook for the Biological Sciences,* Harcourt, Brace and World, New York, 1958.

Weisz, Paul B., *Laboratory Manual in the Science of Biology,* McGraw-Hill, New York, 1959.

PROJECTS

1. Prepare permanent-mount slides by imbedding tissues in paraffin and sectioning with a hand or rotary microtome. Consult a microtechnique book for directions. A good book on microtechnique of plant tissues is *Plant Microtechnique* by Donald A. Johansen (McGraw-Hill, New York, 1940, Chapter 12). For animal tissues try *Animal Micrology* by Michael F. Guyer (The University of Chicago Press, Chicago, Ill., 1953, pp. 35-41, 47-48).
2. Compare cells of other plants and animals with the ones used in this exercise. Look for similarities and differences in structure.

EXERCISE 4
Photosynthesis

MATERIALS

Student Station

test tubes and rack
100-watt lamp
100-ml graduated cylinder
microscope

Each Group of Four Students

100 ml Benedict's solution
100 ml 95% alcohol
50 ml 1:1 ether-benzene solution
100 ml 10% glucose solution
100 ml 10% sodium bicarbonate solution
iodine solution in dropper bottle (as prepared for Exercise 2)
leaves such as spinach, *Tradescantia,* or privet (some leaves from a plant kept in dark for several days)
pith sections
razor blades

microscope slides and coverglasses
several 100-ml beakers
hotplate
filter paper sheets
scissors
ruler
droppers with small tips (or capillary tubes)
4 3-in. sprigs of *Elodea* cut at both ends
mortar and pestle
distilled water

OBJECTIVES

1. To ascertain the site of the photosynthetic reaction.
2. To examine the general nature of the photosynthetic pigments.
3. To demonstrate selected portions of the photosynthetic reaction.

Part 1
Site of Photosynthesis

Photosynthesis is confined to plant cells containing chloroplastids. These cells comprise a small proportion of the total cells that make up a plant. In this part of the exercise you will make leaf cross sections and examine them with the microscope.

Obtain a length of pith and split it lengthwise for about 1 in. Insert a portion of a leaf, and make thin sections as you did in Exercise 3.

Mount your best leaf sections on a microscope slide in a drop of water. Examine the slide under low and high powers, comparing your sections with the diagram in Chapter 4 of *Fundamental Concepts of Biology*.

1. *Which layers of cells appear to contain the most chloroplasts?*

2. *Do the cells comprising the upper and lower surfaces of the leaf contain chloroplasts?*

3. *On which surface of the leaf are the stomata located?*

4. *What is the adaptive advantage of the vertical arrangement of the palisade cells?*

Part 2
Photosynthetic Pigments

The light-sensitive material in leaves is a mixture of pigments rather than a single compound. The technique of paper chromatography can be used to show this. First, a chlorophyll extract is prepared by boiling leaves in alcohol. This is placed on a piece of filter paper. When the paper is exposed to a solvent, the pigments in the extract separate to form bands. These bands have been identified by other techniques as chlorophyll *a* (blue-green band), chlorophyll *b* (yellow-green), xanthophyll (yellow), and carotene (orange).

Place several leaves in a beaker containing a small amount of boiling (95%) alcohol. This should be done over a hotplate rather than an open burner. Remove and save the leaves after they become colorless. Continue boiling the alcohol until a concentrated green soup has been prepared. This is the chlorophyll extract.

Cut a strip of filter paper 1 in. wide and slightly longer than the 100-ml graduated cylinder. Trim one end to a point and pencil a line across the strip 1 in. from the point (see Fig. 4.1).

Using a dropper with a small tip (or a capillary tube), make a narrow band of the extract along the pencil line. Let it dry, then repeat the procedure several times.

Place several milliliters of the ether-benzene solvent in a 100-ml graduated cylinder. Position the strip of filter paper so that the pointed end touches the solvent. The paper should not touch the sides of the cylinder except at the top,

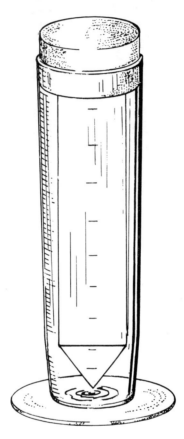

Figure 4.1 Filter paper strip.

where it is held in place with a stopper. In 5 to 10 min your chromatogram should show the bands of pigment described earlier.

Identify each band by its color and note the order of separation on the filter paper.

1. *Which of the pigments is the most soluble in the solvent? What is the evidence for your answer?*

2. *Why are the yellow pigments in leaves usually not visible?*

3. *Xanthophyll and carotene influence the bright colors of leaves in the fall of the year. Why do they only show up then?*

4. *Explain why the term* chlorophyll, *as used in a general sense, denotes a mixture of substances.*

Part 3
The Photosynthetic Reactions

Photosynthesis is an exceedingly complex series of chemical events; consequently, we can examine only selected parts of the reactions.

A. *Products of Photosynthesis*

1. Place a leaf from which the chlorophyll was extracted in Part 2 in a shallow container, and cover it with iodine solution. *Describe the result and its significance.*

2. Obtain a leaf that has been kept in the dark for several days. Place it in boiling alcohol until the chlorophyll is removed, and then test the leaf with iodine solution for the presence of starch. *State your observation and explain its significance.*

3. Grind up several leaves in a mortar with a small amount of warm water. Let the mixture stand for several minutes; then fill a test tube one-half full with the mixture. Add Benedict's solution and place in a boiling water bath. *Record the result and explain its significance.*

4. Obtain another starch-free leaf, place it in a beaker of 10% glucose solution, and return it to the dark for 2 hr. At the end of this time, test the leaf for starch by adding a few drops of iodine solution. *State your result and explain what happened.*

B. *Light Intensity and Photosynthesis*

A sprig of *Elodea* in a tube of water produces bubbles when brightly illuminated. Counting the number of bubbles per unit of time gives a measure of the rate of photosynthesis under these conditions.

Place a 3-in. piece of *Elodea* in a test tube of 10% sodium bicarbonate solution and another sprig in a tube of distilled water. Place both tubes on a rack 6 in. from a 100-watt light bulb. When bubbles seem to be coming off at a steady rate, count the number of bubbles per 5-min interval in each tube. Record your results in Table 4.1. Move the light 3 in. away, wait 5 min, and take another count. Repeat this procedure until you have completed the table.

Table 4.1

Distance from Light	Number of Bubbles per 5 Minutes	
	Distilled H_2O	Carbonated H_2O
6 in.		
9 in.		
12 in.		
15 in.		
18 in.		

Plot both sets of data on the same graph by using a dotted line for distilled water and a solid line for carbonated water.

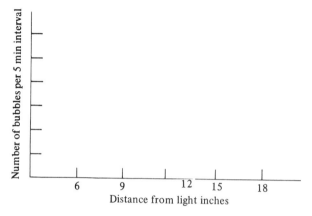

1. *What was the gas in the bubbles given off by the* Elodea?

2. *What does the graph indicate about the effects of light intensity on* Elodea?

3. *What does the graph indicate about the effects of CO_2 on photosynthesis in* Elodea?

SUMMARY

1. The major site of photosynthesis is found in leaves, especially in mesophyll cells.
2. Chlorophyll is a mixture of pigments rather than a single compound. Often the mixture contains chlorophyll *a*, chlorophyll *b*, carotene, and xanthophyll.
3. Starch is found in leaves that have been exposed to light but not in leaves kept in the dark.

4. Leaves will convert glucose to starch in the absence of light.
5. The rate of photosynthesis is proportional to light intensity and amount of CO_2 available.

REFERENCES

Abramoff, Peter, and Robert G. Thomson, *Laboratory Outlines in Biology*, W. H. Freeman and Co., San Francisco, 1963.

Korn, Robert W., and Ellen J. Korn, *Investigations into Biology*, John Wiley and Sons, New York, 1965.

Miller, David F., and Glenn W. Baydes, *Methods and Materials for Teaching the Biological Sciences*, McGraw-Hill, New York, 1962.

Otto, James H., Albert Towle, and Elizabeth H. Crider, *Biology Investigations*, Holt, Rinehart and Winston, New York, 1963.

PROJECTS

1. Make sections of leaves from plants adapted to different habitats. Attempt to correlate their internal structure with the habitat in which they live.
2. Run chromatograms on a variety of leaves and compare their pigments.
3. Using the *Elodea* techniques from Part 3B, test other variables that might influence the rate of photosynthesis; for example, temperature, pH, and different colors of light.

EXERCISE 5
Respiration: Energy Harvest

MATERIALS

Student Station

test tubes and rack
5 test tube respirometers
1 small animal respirometer

Each Group of Four Students

100 ml yeast suspension (½ package yeast/100 ml water)
100 ml distilled water
100 ml 5% glucose solution
100 ml 0.01 molar NaF
100 ml 0.05 molar NaF
100 ml 0.10 molar NaF
50 ml 0.05% methylene blue
50 ml formalin
water bath
several millimeter rulers
frogs or other small animals

OBJECTIVES

1. To investigate the effect of an antimetabolite on cellular respiration.
2. To show that hydrogen is removed from compounds during respiration.
3. To measure the rate of respiration in an organism.

Part 1
Effect of an Antimetabolite on Respiration In Yeast Cells

It is difficult to study or experiment with the complexities of cellular respiration without utilizing sophisticated equipment and techniques. Yeast cells are good laboratory material in this respect, since they provide an easily cultured mass of single cells undergoing respiration. In the following exercise we shall interfere with one step in the yeast's respiratory cycle and observe the consequences.

An enzyme, *enolase,* catalyzes the reaction of PGAL to pyruvate (see page 74 in *Fundamental Concepts of Biology* for a summary of the respiratory cycle). Enolase requires the presence of magnesium ions. In the following experiment, fluoride will be used to precipitate magnesium ions and thus inhibit the reaction. The inhibition is detected as a lowered respiratory rate (CO_2 production) in the yeast cells. Fluorides are also extremely poisonous to humans, so handle the solutions carefully in this exercise and wash your hands immediately afterward.

To construct a simple respirometer fill one of the small, flat-bottomed culture tubes with water (yeast suspension in an actual experiment). Slide a large test tube down over the culture tube as far as it will go, as shown in in Fig. 5.1. Invert the

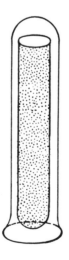

Figure 5.1

tubes quickly, so that very little fluid is lost from the culture tube. The air space that appears in the culture tube is measured with a millimeter ruler. If the culture tube contains respiring yeast cell, CO_2 will accumulate to displace additional fluid. After a period of time, the size of the air space provides a measure of the amount of respiration that has taken place.

Label culture tubes 1 through 5, holding each tube upside down as you apply the number. Fill each tube as follows:

Tube 1 5 ml yeast suspension; 10 ml distilled water
Tube 2 5 ml yeast suspension; 5 ml glucose solution; 5 ml distilled water
Tube 3 5 ml yeast suspension; 5 ml glucose solution; 5 ml 0.01 M NaF
Tube 4 5 ml yeast suspension; 5 ml glucose solution; 5 ml 0.05 M NaF
Tube 5 5 ml yeast suspension; 5 ml glucose solution; 5 ml 0.10 M NaF

Invert each culture tube in a large tube as illustrated previously. Measure the air space in each of the respirometers and record it below. Place the respirometers in a 37° C water bath for 1 hr. Measure the air space again and record under Final Reading.

Tube Number	Initial Reading	Final Reading	Difference
1			
2			
3			
4			
5			

1. *What was the function of Tube 1?*

2. *Why was glucose necessary in Tubes 2 to 5?*

3. *Which tube showed the highest rate of respiration?*

4. Which tube showed the lowest amount of respiration?

5. Explain whether the yeast cells were respiring aerobically or anaerobically in this experiment.

Part 2
The Removal of Hydrogen (Dehydrogenation) during Respiration

Methylene blue, an organic compound, is a hydrogen acceptor. It combines with hydrogens that are removed from other compounds, for example, during cellular respiration. As it takes on hydrogen it becomes colorless, hence the color change shows that hydrogen is being produced in a reaction.

Number a set of test tubes 1 to 5 and fill them as directed below.

Tube 1 5 ml yeast suspension; 10 ml distilled water
Tube 2 5 ml yeast suspension; 5 ml glucose solution; 5 ml distilled water
Tube 3 5 ml yeast suspension; 5 ml glucose solution; 5 ml 0.01 M NaF
Tube 4 5 ml yeast suspension; 5 ml glucose solution; 5 ml 0.10 M NaF
Tube 5 5 ml yeast suspension; 5 ml glucose solution; 5 ml formalin

Add to each tube 0.05% methylene blue drop by drop, stirring frequently until a light blue color persists in the tube. Seal the tubes with stoppers, note the time, and put them in a test tube rack. Record the time at which the color disappears from each tube.

Tube Number	Starting Time	Time when Color Changed	Time Difference
1			
2			
3			
4			
5			

1. *Why will Tube 5 never become colorless?*

2. *Which tube changed first? Explain why.*

3. *Why was Tube 1 included in the experiment?*

Part 3
Measuring the Respiratory Rate in Organisms

---

1. The Warburg respirometer.

The standard method for accurately measuring the respiratory rates of small organisms involves the use of the Warburg apparatus. See Fig. 5.2. Although the apparatus looks complex, it consists of three basic parts: a water bath (the largest part of the apparatus) in which temperature is maintained within narrow limits; small glass vessels which contain the organisms being studied; and a U-shaped tube (manometer) attached to each glass vessel. See Figure 5.3. The manometer is graduated and is filled with a special fluid just prior to being used. One arm of the manometer is open and thus allows the fluid to respond to atmospheric pressure.

In very general terms, the apparatus is operated as follows. Organisms are placed in the small glass vessels which are, in turn, attached to the manometers. The vessels are then immersed in the constant temperature water bath. After about 20 min. the vessels are sealed from the atmosphere by closing their glass valves. The organisms in the vessels then begin to use the air (oxygen) in their surroundings. This is indicated by movement of the fluid in the manometers. Manometer readings are made at frequent intervals over a specified time interval. The difference between the initial reading and the last reading indicates the volume of oxygen that has been consumed. By the use of appropriate calculations, one can convert the manometer volumes into the actual amounts of oxygen used by the organisms during a period of time. This is termed the organism's rate of respiration.

As with any precision apparatus, certain procedures must be followed carefully to get reliable results. For example, during the course of an experiment, one empty vessel and its manometer must be included in the test run. The manometer on this empty vessel detects changes in atmospheric pressure that may occur during the experiment. Such changes affect all of the manometers and must be included in their readings.

Since organisms give off carbon dioxide during their respiration, this gas must be disposed of so that it will not interfere with the oxygen volume determination. Usually this is accomplished by including sodium hydroxide in the glass vessels with the organisms. The expired carbon dioxide reacts with the sodium hydroxide and thus does not remain as a gas in the system.

In the next exercise you will assemble and operate a simple respirometer based on the principle of the Warburg apparatus.

Figure 5.2 Warburg apparatus (Courtesy American Instrument Company, Division of Travenol Laboratories, Inc.).

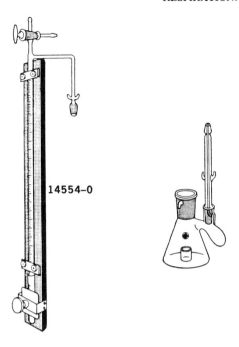

Figure 5.3 Manometer and glass vessel.

2. A simple laboratory respirometer.

Inexpensive respirometers can be constructed with simple laboratory materials, as we shall show. These are not as accurate or precise as a research instrument like the Warburg respirometer, but nevertheless they are satisfactory if operated properly.

Assemble the apparatus shown in Fig. 5.4. When an animal is sealed in the respiration chamber, it uses up oxygen and gives off carbon dioxide. The carbon dioxide is absorbed by sodium hydroxide. As the oxygen is used, the change in pressure is registered in the fluid in the manometer tube. The water bath holds the apparatus at a constant temperature while the experiment is in progress. By noting the volume of oxygen used per unit of time and knowing the weight of the organism, a simple calculation gives the animal's metabolic rate.

Place a frog (or other small animal) in the respiration chamber bottle with all vent tubes open. After 5 min close the vent tubes, record the time, and take a reading on the manometer tube. At the end of a 15-min period, take a second reading on the manometer tube. Open the vent tubes, let the manometer equalize, then repeat the procedure for a second measurement. Take at least three measurements in this way (they should be reasonably similar), then use the average of the three for calculating the animal's metabolic rate.

To calculate the metabolic rate, convert the milliliters of oxygen per 15-min period to milliliters per hour. For example,

$$\frac{5 \text{ ml O}_2}{15 \text{ min}} = \frac{X \text{ ml O}_2}{60 \text{ min}}$$

$$X = \frac{(60)(5)}{15} = 20 \text{ ml O}_2 \text{ per hour}$$

Weigh the animal you used, then divide the oxygen rate per hour by the weight to obtain milliliters O_2 per hour per gram of body weight. This is the metabolic rate of your organism.

Repeat this experiment with a frog that has been kept in a refrigerator.

1. *During the experiment, why was it necessary to open the vents and equalize the manometer between each measurement?*

2. *Why is it necessary to use a carbon dioxide absorbing material during the experiment?*

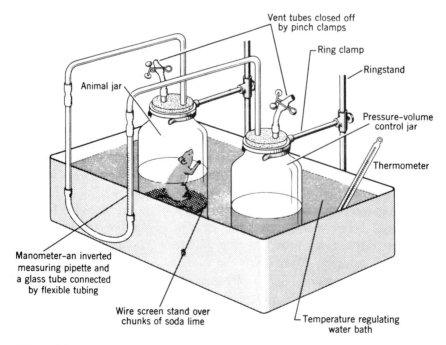

Figure 5.4 A respirometer for measuring oxygen consumption in small animals.

3. *How would you prepare a control experiment to run with this exercise?*

4. *How did the metabolic rate of the refrigerated frog compare with that of the room temperature frog? Account for the difference.*

5. *Would a warm-blooded animal like a mouse give comparable results in a respirometer?*

SUMMARY

1. Cellular respiration is inhibited if one of the links in the chains of chemical reactions is broken. An antimetabolite is a material that interferes with the sequence of vital chemical reactions in an organism.
2. Methylene blue, a hydrogen acceptor, provides a simple technique for showing that dehydrogenation occurs during respiration. It can also be used to indicate that respiration has decreased in rate or has ceased to occur. Yeast cells provide a useful organism for respiration experiments.
3. The respiratory rate and metabolic rate can be determined with a relatively simple respirometer. Metabolic rate in cold-blooded animals is closely related to the temperature of their environment.

REFERENCES

Humphrey, Donald G., Henry Van Dyke, and David L. Willis, *Life in the Laboratory,* Harcourt, Brace and World, New York, 1965.

Korn, Robert W., and Ellen J. Korn, *Investigations into Biology,* John Wiley and Sons, New York, 1965.

Weisz, Paul B., *Laboratory Manual in the Science of Biology,* McGraw-Hill, New York, 1963.

PROJECTS

1. Using a respirometer, calculate the metabolic rate of more than one of the same kind of animal and determine whether or not there is a characteristic metabolic rate for the species.
2. Determine the metabolic rates of a series of different animals.
3. Attempt to establish a relationship between body size and respiratory rate.
4. Determine the respiratory rates on an organism at different temperatures. Graph your results and see if you can derive a generalization concerning the effects of temperature on respiration.
5. Ascertain the effects of certain hormones and drugs on the metabolic rate of a frog.

EXERCISE 6
Transport of Materials

MATERIALS

Student Station

dissecting instruments (needle, probe, scalpel, scissors, pins)
small cork
droppers
150-ml beaker
celery petiole
razor blade

Each Group of Four Students

frogs
0.7% saline solution
hotplate
ice
4 250-ml beakers
3 in. × 5 in. cardboard with hole ½ in. in diameter
compound microscope

64 EXPERIMENTS IN FUNDAMENTAL CONCEPTS OF BIOLOGY

radish seedlings grown between moist filter paper
microscope slides and coverglasses
1% aqueous eosin solution
3 500-ml flasks with side arm
3 2-hole rubber stoppers to fit 500-ml flask
3 pieces of 4-in. glass tubing to fit stoppers
6 small pieces of rubber tubing
3 small funnels
3 pinch clamps
3 1-ml graduated pipettes
3 ringstands with clamps
stems and leaves of bean or similar plant cut under water and held in beaker of water (Azalea, if available, makes excellent material)
petroleum jelly
plastic bag
small fan (optional)

OBJECTIVES

To study the structural adaptations in selected plants and vertebrate animals in relation to the problem of transport of materials.

**Part 1
The Frog**

Pith a frog according to directions on p. 199. Place frog dorsal side down in a dissecting pan. Make incisions to expose the body cavity. Cut the corners of the lower jaw so that it can be held or pinned down out of the way. Keep all of the exposed surfaces moist by adding drops of saline solution as needed.

Examine the dissected frog closely. Notice the teeth on the upper jaw and the attachment of the tongue.

1. *Frogs do not chew their food. Of what significance are the two kinds of teeth?*

2. *Where is the tongue attached?*

3. *Of what adaptive value is this?*

Place a very small piece of cork on the palate between and just posterior to the eyes. Observe the movement of the cork and relate this to the movement of food. Record the time it takes the cork to move 1 cm.

1. *What might account for this movement?*

2. *How might you test your hypothesis? (Check a reference book to see if you are right.)*

3. *Why do you suppose the cork moves into the esophagus and not away from it?*

Place a few drops of *warm* saline solution on the palate. Record the time it takes the cork to move 1 cm. Now place a few drops of *cold* saline solution on the palate and record the time it takes the cork to move 1 cm.

	Time per centimeter
Warm saline	
Cold saline	

1. *What is the relationship between temperature and movement of the cork?*

2. *Frogs do not contain any internal devices for maintaining constant body temperature. Do your observations suggest any possible connection between movement of food and the time of the year frogs are most active?*

Continue to dissect the frog. If your frog is a female, the swollen ovary with the mass of eggs will interfere with observations. Remove the eggs at this time. Locate and cut open the stomach. Locate also the following structures: small and large intestines, cloaca, cloacal opening, liver, heart, and lungs (Fig. 6.1)

Food passes from the esophagus into the stomach. Note the numerous folds in the stomach lining. The movement of food into the small intestine is controlled by a sphincter muscle, the pylorus, which can be distinguished at the posterior end of the stomach by a constriction. Note the numerous blood vessels surrounding the intestine. Digested food moves into these blood vessels and is carried to all parts of the body.

Trace the pathway of waste materials through the small intestine to the large intestine and cloacal area. Waste materials are released through the cloacal opening.

1. *What is the significance of numerous folds in the inner lining of the stomach?*

2. *Why is the small intestine many times longer than the large intestine?*

TRANSPORT OF MATERIALS   67

3. *Why do you suppose there is a blood vessel from the small intestine to the liver?*

Locate the lungs in the anterior portion of the body cavity. They are usually inflated. Note that the lungs are supplied with blood vessels. In frogs, skin is also involved in respiration.

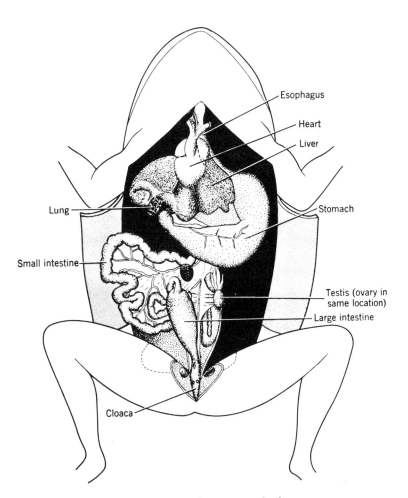

Figure 6.1 Internal anatomy of a frog.

Food and gases are carried throughout the body in the blood stream. It is possible to demonstrate the movement of blood by inspecting the web between the toes with the low power of the microscope.

Spread the toes apart and pin one foot to a piece of cardboard with a hole in it (see Fig. 6.2). Place this on the microscope and, with the low power, observe the flow of blood in the capillaries and in other small blood vessels.

1. *How does the size of the blood cells compare with the size of the capillaries?*

2. *Of what adaptive significance is this?*

3. *What major event important in transport occurs in the capillaries?*

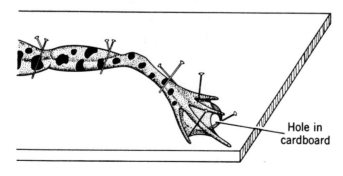

Figure 6.2 Blood flow in the web of a frog's foot.

TRANSPORT OF MATERIALS

Part 2
Transport in Plants

The transport system of plants is composed of water-conducting xylem and food-conducting phloem cells. See *Fundamental Concepts of Biology*. The following experiments are concerned with the problem of water movement.

A. Root Hairs (Uptake of Water)

The only part of plant roots involved with the uptake of water and minerals are root hairs. Observe root hairs of radish seedlings which have been grown in petri dishes on filter paper. With the aid of a hand lens or the low power of the microscope, observe the root hairs closely.

1. *Describe the appearance and growth habit of the root hairs.*

2. *What adaptive value is there to the presence of many small root hairs rather than a few large root hairs?*

3. *Why is it that when a small tree is transplanted, it is necessary to keep the soil well watered for a short period of time?*

B. Movement of Water

Once water and minerals enter the plant they move into the xylem (see *Fundamental Concepts of Biology* and pass up through the roots and stems into the leaves. The movement of water can be observed in the celery petiole.

Place a petiole in a 150-ml beaker containing 50 ml of 1% eosin solution. Observe the movement of the colored water. Set aside for 1 hr and go on to Part C before continuing with this experiment.

After 1 hr cut a thin cross section of the petiole with a razor blade and mount on a slide. Observe with the low and high powers of the microscope.

1. *Is the dye distributed throughout the cross section or is it concentrated in certain places?*

2. *How do the cells surrounding the xylem receive water?*

C. Release of Water

Loss of water—transpiration—may be detrimental to a plant if the amount of water lost seriously exceeds the amount of water absorbed. However, transpiration plays an important part in moving water, and during the process some loss of water is essential. A simple potometer will serve to demonstrate the phenomenon of transpiration by measuring the amount of water absorbed.

Set up three potometers as in Fig. 6.3. Fill a 500-ml side arm flask with water. Plug the flask with a two-hole rubber stopper which has a 4-in. piece of glass tubing through one of the holes. Attach a small piece of rubber tubing to the glass tubing above the stopper. Insert a small funnel at the top of the rubber tubing. A pinch clamp should be placed on the rubber tubing. Fit a small piece of rubber tubing on the side arm of the flask and insert a 1-ml graduated pipette into the rubber tubing. A ringstand and a clamp will serve to hold the pipette in a horizontal position. Remove a plant stem from the beaker of water and quickly force it through the second hole in the rubber stopper. It must fit snugly. If it does not, coat the hole with petroleum jelly to stop air leaks. Fill the flask and pipette completely by adding water to the funnel. Tip the apparatus slightly and allow a few drops to drip from the end of the pipette. Release the pinch clamp. Water should not continue to drip from the end of the pipette. If it does, check for air leaks. Coat rubber-to-glass connections until leak stops. Place a plastic bag over the plant in one setup; expose the second setup to a breeze from an open window or small fan; and allow the third setup to remain at room conditions.

Allow the setups to stabilize for 5 min at the various experimental conditions. Fill the pipettes as before. Record the amount of water absorbed at intervals of 30, 60, 90, 120, 150, and 180 sec by recording the amount of water lost from the pipette.

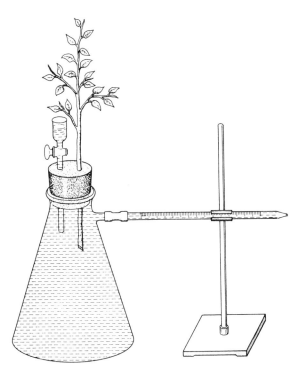

Figure 6.3 Potometer.

Experiment 1 (plastic bag)

Pipette reading
    (water lost) ―――――――――――――――――――
                 30 60 90 120 150 180 sec

Experiment 2 (wind)

Pipette reading
    (water lost) ―――――――――――――――――――
                 30 60 90 120 150 180 sec

Experiment 3 (room conditions)

Pipette reading
    (water lost) ―――――――――――――――――――
                 30 60 90 120 150 180 sec

1. *In which experiment did the fastest water movement occur?*

2. *In which experiment did the slowest water movement occur?*

3. *Does this give you some clue to the reasons why plants require more water on some days than on others?*

4. *Can you think of another environmental factor in addition to moisture and air circulation that will determine a plant's requirement for water at a particular time?*

5. *What adaptations would you expect a plant to have for existence in an area exposed to hot, dry winds?*

6. *What other factors in addition to transpiration account for movement of water up through a plant?*

SUMMARY

The frog is adapted to its way of life, as shown by the way in which it captures food and the structures it has for digesting food and eliminating waste. Its circulatory system is efficient in moving nutrients to all parts of the body and removing waste products of cellular metabolism. Exchange of gases occurs both in the lungs and through the skin. The absence of mechanisms for controlling internal body temperature limits the frog's activity to certain seasons of the year.

A plant, compared to a frog, has simple but efficient adaptations for transport of water. Transpiration plays an important role in moving water through plants. Water enters root hairs, moves into xylem, and is conducted through the xylem into the leaves. The adaptations for controlling water loss due to transpiration are closely correlated to the physical environment.

REFERENCES

Humphrey, Donald G., Henry Van Dyke, and David L. Willis, *Life in the Laboratory*, Harcourt, Brace and World, New York, 1965.

Lawson, Chester A., Ralph W. Lewis, Mary Alice Burmester, and Garrett Hardin, *Laboratory Studies in Biology*, W. H. Freeman and Co., San Francisco, 1955.

PROJECTS

1. Observe feeding in single-celled organisms such as protozoa. Note that even though there is a lack of a complex transport system, there are adaptations for movement of substances within the organism. [*Note:* Turtox supplies *Didinium* in the same culture with *Paramecium,* upon which it feeds. Also, Southern Biological Supply Co. (McKenzie, Tennessee, 38201) has a number of protozoa cultures supplied with organisms upon which each feeds.]
2. Make a comparative study of mechanisms for transport. Use organisms such as *Hydra* and its gastrovascular cavity, planaria and its branched digestive tract and one opening, and an earthworm and its well developed digestive tract with a mouth and anus. Why is it impossible for a planarian worm to be round?
3. Study root systems of a carrot, a water hyacinth or similar water plant, a cactus, and a bromeliad. How is each adapted to its particular way of life?

EXERCISE 7
Control Within Cells

MATERIALS

Each Group of Four Students

styrofoam cut according to instructions on p. 64
4 scalpels
pipe cleaners
4 test tubes in rack
10 ml 95% cold alcohol
glass rods with hooks at tips

Class Stock

50 g liver minced in blender and poured into beaker with 100 ml 5% sodium lauryl sulfate solution. The mixture should be stirred gently for 1 hr and then centrifuged for a short time.

OBJECTIVES

1. To construct a model of a DNA molecule from styrofoam.
2. To examine crude DNA extracted from living cells.

Part 1
A Styrofoam Model of DNA

Biology and other sciences often use models to illustrate concepts or structures that are not easily visualized. It should be kept in mind that most models are not designed to be accurate representations but rather are intended as analogous objects. Watson and Crick, for example, used pieces of tin to construct a DNA model. This was a hypothetical model based on their X-ray diffraction studies of DNA.

Models can be extremely useful in indicating relationships between parts, in showing how an object functions, and even for performing hypothetical experimental procedures. For example, one can manipulate parts of a model and theorize about the effects when such an activity might be difficult or impossible to do with the actual material.

To make the model. Take a strip of styrofoam and use the forms shown as patterns.

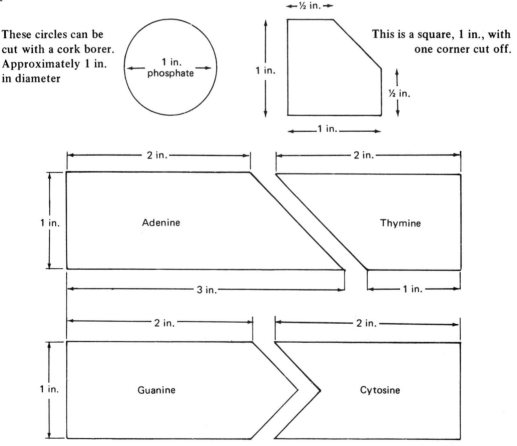

These can be cut as strips 4 in. x 1 in. and then cut into two pieces as shown.

CONTROL WITHIN CELLS 77

Cut about 16 copies of each pattern for each group. They can be attached to one another with pipecleaners.

Start by assembling a guanine *nucleotide* (5 carbon sugar + phosphate group + guanine) as follows:

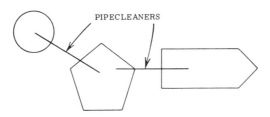

Now assemble the complementary nucleotide, cytosine, which should resemble this:

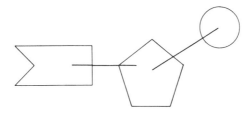

Assemble an adenine and a thymine nucleotide. They should look like this:

Assemble your four nucleotides into a single chain to look like the one shown on page 78.

Add four additional nucleotides to the end of the chain in the following order: cytosine, cytosine, adenine, thymine.

You now have a strip of eight nucleotides. Assemble the *complementary* strip that is needed to complete the model. Use short pieces of pipecleaners as hydrogen bonds to link complementary nucleotides. When completed, you have a crude model of a segment of a DNA molecule. If you can visualize many thousands of nucleotides in this kind of chain, the immense size of this molecule begins to become evident.

A double helix effect can be approximated by linking your model with another one and gently twisting them.

Using your model for a guide, answer the following questions.

78   EXPERIMENTS IN FUNDAMENTAL CONCEPTS OF BIOLOGY

1. *In what sense is DNA a monotonous (repetitive) molecule?*

2. *Which part of your model is analogous to coded information?*

3. *What links adjacent phosphate molecules?*

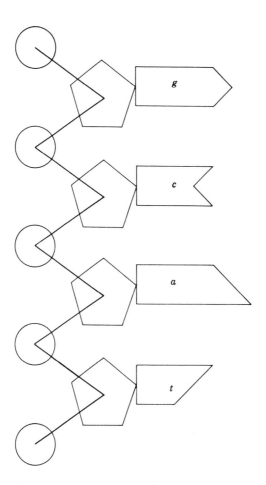

CONTROL WITHIN CELLS

Messenger RNA

Messenger RNA is composed of a single strand of nucleotides with uracil occurring in place of thymine. To get an idea of how mRNA is formed, use the following procedure.

Lay your DNA model on a flat surface. Gently open the hydrogen bonds between the last six nucleotides so that the two nucleotide chains can be separated as illustrated. Use one of the chains as a template to assemble a six-nucleotide RNA molecule.

1. Would the RNA be the same if copied from either chain? Why?

2. What happens to this RNA after being assembled in the nucleus?

3. What do you suppose happens to the DNA after the RNA leaves it?

4. Where did the nucleotide units that were used in making RNA come from?

5. *The mRNA that you assembled could theoretically serve as a template for the linking of how many amino acid units carried by transfer RNA?*

6. *If your mRNA model reads adenine-uracil-cytosine-cytosine-adenine-uracil, what is the sequence in the two tRNAs required to match it properly?*

7. *How does this tRNA sequence compare with the DNA sequence of bases from which you assembled the messenger RNA model?*

Part 2
Crude DNA from Living Cells

Pour 2 ml of the supernatant of the liver extract in a test tube. Slowly add 5 ml of 95% alcohol which has been kept in a deep freeze. If this is done properly, you will have a layer of alcohol above the extract.

Insert a glass rod with a hook into the extract and pull it up into the cold alcohol layer. A small fiber should form on the glass hook. By repeating this procedure, a small mass of fibers can be extracted. These fibers contain a mixture of DNA, RNA, and other organic materials, but they have at least a general resemblance to purified DNA.

1. *How would you describe the consistency of the fibers?*

2. *What does this indicate about the structure of the DNA molecule?*

3. *In procedures like this why would you nearly always end up with a mixture rather than pure extracts?*

SUMMARY

1. A model of DNA constructed from styrofoam pieces illustrates the basic building units of DNA: sugars, phosphates, and nitrogen bases. Nucleotide units can be linked in various sequences to illustrate the coded information concept of DNA. Messenger RNA is synthesized under the direction of a segment of DNA.
2. Crude DNA can be extracted from cells in the form of viscous fibers.

REFERENCES

Weisz, Paul B., *Laboratory Manual in the Science of Biology,* McGraw-Hill, New York, 1963.

Prepared DNA models are available as supplementary materials from the following sources:

DNA Model Kit (paper). Burgess Publishing Co., 426 South 6th Street, Minneapolis 15, Minn.

DNA Kit for Student Use (plastic pieces). KD Biographics, 1050 Flake Drive, Palatine, Ill., 60067.

DNA: The Molecule of Life (plastic tubing). Silver Burdett Company, Morristown, N. J.

Ward's Natural Science Establishment, Inc., P. O. Box 1712, Rochester, N. Y., 14603.

Turtox Products, CCM: General Biological, Inc., 8200 South Hoyne Avenue, Chicago, Ill., 60620.

EXERCISE 8
Control By Chemical Agents

MATERIALS

Each Group of Four Students

4 small tomato plants
lanolin (sheep fat)
lanolin containing 1% indoleacetic acid
millimeter ruler
scalpel or razor blade
labels
3 2-day old Leghorn male chickens
chicken feed
cages for chickens
1 ml syringe graduated in tenths
20-gauge hypodermic needle
30 mg testosterone propionate
sesame oil

84   EXPERIMENTS IN FUNDAMENTAL CONCEPTS OF BIOLOGY

OBJECTIVES

1. To investigate the relationship of auxin to apical bud dominance and inhibition of axillary bud development.
2. To test the effect of testosterone on comb formation in chickens.

**Part 1
Apical Bud Dominance**

Young tomato plants with at least three axillary buds and no evidence of branching are satisfactory for this experiment. (See Fig. 8.1.) Select four plants and label them 1 through 4. Treat as follows:

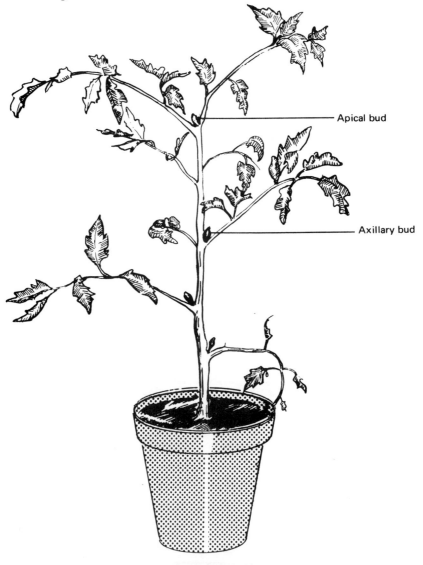

Figure 8.1   Tomato plant showing axillary and apical buds.

CONTROL BY CHEMICAL AGENTS   85

Plant 1   Control. Do nothing to this plant.
Plant 2   Cut off the stem just below the apex.
Plant 3   Cut the stem just below the apex and add lanolin to the cut end.
Plant 4   Cut the stem in the same way as for plants 2 and 3 and add lanolin containing 1% indoleacetic acid to the cut end.

With a millimeter ruler measure the length of the first three axillary buds below the apex on plants 1 through 4. Record this information on Table 8.1. Do the same at the end of 1 week and at the end of 2 weeks. (If possible, additional measurements should be taken at intervals of ½ week.)

1. *In which plants was there pronounced growth of axillary buds?*

2. *In which plants was there little or no growth of axillary buds?*

3. *What effect does auxin from the apical bud have on growth of axillary buds?*

4. *What happens if the apical bud is removed?*

5. *Suggest a possible practical application of the information learned from this exercise.*

Table 8.1

Date	Plant 1 length in mm			Plant 2 length in mm			Plant 3 length in mm			Plant 4 length in mm		
	Bud 1	Bud 2	Bud 3	Bud 1	Bud 2	Bud 3	Bud 1	Bud 2	Bud 3	Bud 1	Bud 2	Bud 3
Total growth												

Part 2
The Effect of Testosterone on Comb Formation in Chickens

The following exercise illustrates the action of testosterone in producing one of the secondary sex characteristics in the male chicken, namely, comb formation.

Obtain three 2-day old male Leghorn chickens. One serves as a control and will receive no treatment other than sufficient food and water. Treat the other two as follows: At 2-day intervals for 2 weeks inject one chicken with 5 mg of testosterone propionate in 0.1 ml of sesame oil at each injection. At the same intervals, inject the other chicken with 0.1 ml sesame oil at each injection. The easiest place to inject is at the back of the neck. Pull the skin gently and inject the testosterone just below the surface. Observe closely for 2 weeks. Feed the experimental chickens in the same way as the control. Record your observations in Table 8.2.

1. *Is there any difference in the combs and wattles of the control chicken and the one treated with sesame oil?*

2. *Is there any difference between the chicken treated with sesame oil and the one treated with testosterone and sesame oil? Describe.*

CONTROL BY CHEMICAL AGENTS 87

	Table 8.2A. Week One		
	Injection Dates and Comments		
Treatment	1	2	3
Testosterone and sesame oil			
Sesame oil			
Control			

	Table 8.2B Week Two		
	Injection Dates and Comments		
Treatment	1	2	3
Testosterone and sesame oil			
Sesame oil			
Control			

3. *Is there any difference between the control and the chickens treated with sesame oil and testosterone? Describe.*

4. *What was the reason for treating one of the chickens with sesame oil without testosterone?*

5. *Did you note any effects of testosterone other than its effect on the combs and wattles? Describe.*

6. *Why is it important to feed the experimental chickens in the same way as the control?*

SUMMARY

1. The apical bud of plants produces auxin, which inhibits growth of axillary buds.
2. Testosterone influences secondary sex characteristics.

REFERENCES

Feldman, Solomon, *Techniques and Investigations in the Life Sciences,* Holt, Rinehart and Winston, New York, 1962.
Newcomb, Eldon H., Gerald C. Gerloff, and William F. Whittingham, *Laboratory Studies in Biology,* W. H. Freeman and Co., San Francisco, 1964.
Weisz, Paul B., *Laboratory Manual in the Science of Biology,* McGraw-Hill, New York, 1963.

PROJECTS

1. Induce rapid metamorphosis in tadpoles with thyroxine. See Turtox Service leaflet number 54 for information.
2. Induce ovulation in frogs by the injection of macerated pituitary glands. Materials and directions are available from biological supply houses.
3. Treat various plant cuttings with a root-forming hormone.
4. Treat oat coleoptiles in the following ways: (*a*) decapitate; (*b*) decapitate, place agar block on cut tip; (*c*) decapitate, place agar block containing auxin on cut tip; (*d*) measure elongation in each over a period of time. For directions see Feldman, Solomon, *Techniques and Investigations in the Life Sciences,* Holt, Rinehart and Winston, New York, 1962, page 39.

EXERCISE 9
Control by Nervous Systems

MATERIALS

Each Group of Four Students

20 wood rectangles about 1 in. X 2 in. X ¼ in. or light weight dominoes
roll of tape
scissors
20 small springs (discarded ball-point pens are a good source)
2 12-in. strips of wood

Class Stock

charts or models of invertebrate nervous systems (see references)
charts or models of vertebrate brains (see references)
chart or models of a mammalian nervous system (see references)

OBJECTIVES

1. To examine, by means of a mechanical model, certain features of the conduction of nerve impulses.

90 EXPERIMENTS IN FUNDAMENTAL CONCEPTS OF BIOLOGY

2. To compare nervous systems of a series of selected organisms.
3. To review the general structure of the mammalian nervous system.

Part 1
The Nerve Impulse: An Analogy from a Mechanical Model of Nerve Conduction

---

Use tape to hinge ten wood rectangles to a strip of wood (or a table top), as shown in Fig. 9.1. The rectangles should be arranged so that each can knock down the one in back of it. When operating properly, the first piece is flipped down and starts a wave of motion that goes to the end of the board. The springs restore the pieces to their upright positions as the wave passes.

Operate the model several times, noting how the wave of activity travels through the model. Assume that this wave of motion is analogous to an impulse traveling along a nerve fiber.

1. *What event in the model is comparable to depolarization?*

2. *What represents repolarization?*

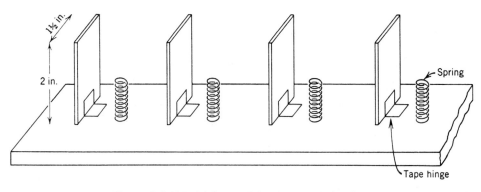

Figure 9.1 Model for studying nerve conduction.

3. *Can self-propagation of an impulse be demonstrated with the model? Explain.*

4. *In what way can the model be used to demonstrate the all-or-none concept?*

5. *Show how there is a threshold strength necessary to start an impulse in the model.*

6. *If you initiate impulses in the model by using different stimuli (like blowing and tapping with a pencil), does it change the nature of the impulse? Does this also apply to a true nerve impulse?*

7. *Verify that the model will conduct an impulse in only one direction. This is also true of a nerve fiber but for a different reason. Explain.*

Part 2
A Comparison of Nervous Systems

Obtain a model or chart that shows the nervous system of an invertebrate such as a crayfish, grasshopper, or earthworm. Preserved specimens of these animals can also be used. For comparative purposes, use a model or chart of a vertebrate that also shows the nervous system.

Observe on the model, chart, or specimen that the invertebrate nervous system is basically a chain of paired ganglia extending the length of the underside of the animal's body. Each ganglion is a cluster of nerve cell nuclei which, with its motor and sensory fibers, innervates the structures within the segment in which it occurs. Neural coordination between different segments occurs by nerve fibers which interconnect adjacent ganglia. These nerves enable an impulse to travel the length of the animal.

The invertebrate brain is not equivalent to the brain of a vertebrate in terms of anatomical or functional complexity. It consists of the ganglia of the head region fused into two masses surrounding the esophagus. It does not show the anatomical specialization into distinct parts that is so characteristic of vertebrate brains. There is also far less functional specialization. For example, removing an insect's brain has little effect on its motor activities.

It would be a mistake, however, to assume that this anatomically simple nervous system is capable of controlling simple activities only. Motor actions like walking, flying, and eating are complex. Also, many invertebrates such as the social insects have life cycles that require a large amount of neuromuscular coordination and behavioral complexity.

After looking at the invertebrate system, examine a model or chart of a vertebrate nervous system. Answer the following questions.

1. *How does the location of the vertebrate system in the body differ from that of the invertebrate?*

2. *How does protection for the systems differ in the two groups?*

3. *Does the vertebrate brain appear to be more complex than the invertebrate brain? In what way?*

4. *Which brain seems larger in relation to the remainder of the nervous system?*

5. *Up to this point we have emphasized differences between the nervous systems of invertebrates and vertebrates. Can you think of any structural components that are alike in the two systems? Are there functional similarities?*

Examine charts, models, or specimens of several vertebrate brains such as fish, amphibian, bird, and mammal. (See Fig. 9.2.) On each one locate the olfactory bulbs, medulla, cerebellum, midbrain, optic lobes, cerebrum, and cranial nerves.

1. *How do the olfactory bulbs of a fish (shark) and a mammal compare? The fish is said to have a "smell" brain. Do you see why?*

2. *Which of the brains has the largest optic lobes? How is this related to the animal's means of survival or food gathering?*

3. *Observe the medulla (brain stem). Is it larger or smaller in relation to the rest of the brain when one compares the lower vertebrates to the higher vertebrates?*

4. *How does the surface of the mammalian cerebellum (motor area) differ from the cerebellum of other vertebrates? What is accomplished by this arrangement? Can you think of a reason why a mammal requires a more complex cerebellum than other vertebrates?*

5. *What happens to the cerebrum as one proceeds from fishes to mammals? In mammals it is sometimes called the neopallium (new cloak). Do you see why?*

6. *From your comparisons, explain why the form of brains support the biological principle that there is an intimate relation between form and function.*

Part 3
The General Structure of the Mammalian Nervous System

Use a chart or model of a mammalian nervous system for the following exercise. Beginning with the brain, note the relative size and complexity of the brain compared with the spinal cord. Observe that the cerebral hemispheres envelop nearly all of the brain except the cerebellum and brain stem, and that the surface of these hemispheres is convoluted. It should be evident that all sense organs in the head region are served by the cranial nerves, which enter and leave the brain. These are homologous to the peripheral nerves, which connect with the spinal cord.

Observe that spinal nerves arise from the spinal cord at regular intervals but are larger in the areas of the forelimbs and hind limbs than elsewhere. The networks of nerves in the limb areas are called plexuses. The spinal nerves serve the skeletal muscles and skin and contain both sensory and motor neurons.

The internal organs are innervated by the autonomic nerves. Locate this part of the peripheral nervous system as a chain of ganglia and small nerve networks near the backbone but on its underside. These autonomic fibers join the other peripheral nerves just before they connect with the spinal cord.

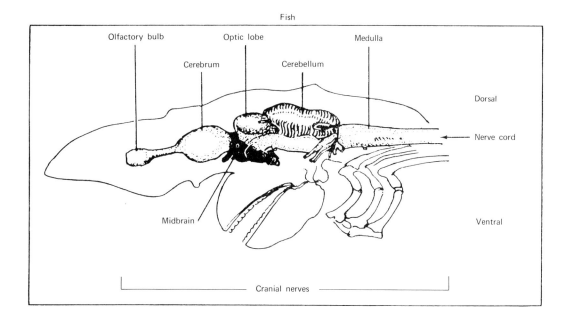
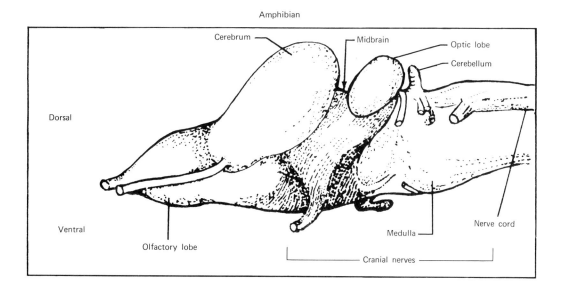

Figure 9.2 Vertebrate brains.

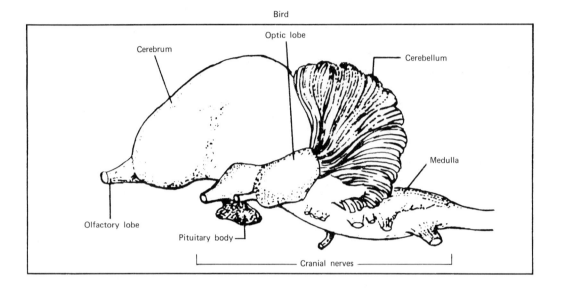

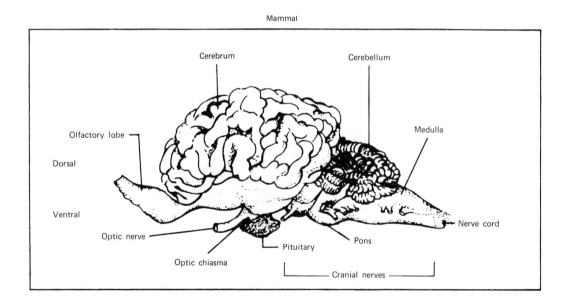

Figure 9.2 Vertebrate brains (continued).

From your observations and with the aid of *Fundamental Concepts of Biology*, answer these questions.

1. *In what way, if any, does the mammalian brain illustrate cephalization?*

2. *Based on its size, what part of the mammalian brain dominates the other parts? Do you think this also holds true from a functional viewpoint? Explain.*

3. *Would you expect most cranial nerves to contain both motor and sensory neurons? Can you think of any cranial nerves that might be entirely sensory?*

4. *What major functions would you give for the spinal cord, considering its location and relation to other parts of the nervous system?*

5. *Why do many internal organs receive fibers from different parts of the autonomic system?*

SUMMARY

1. A mechanical model can illustrate, at least crudely, some aspects of nerve impulse conduction. Depolarization, repolarization, self-propagation, all-or-none concept, threshold, summation, and one-way conduction can be shown by analogy on a model.

2. A comparison of invertebrate and vertebrate nervous systems shows numerous anatomical dissimilarities. Both, however, are composed of neurons, and both act to coordinate complex actions.
3. The mammalian nervous system is exceedingly complex, both anatomically and functionally. It is convenient to refer to the central, peripheral, and autonomic systems, but all are parts of the same structural and functional unit.

REFERENCES

Lavoie, Marcel E., "A Mechanical Model of Nerve Conduction," *Turtox News,* **38** (12), 298-299 (1960).

CCM: General Biological, Inc., Chicago. Jewell models of vertebrate brains. Also a variety of charts.

Ward's Natural Science Establishment, Inc., New York. Models of the vertebrate brain, charts.

EXERCISE 10
The Interaction of Control Systems: Homeostasis

MATERIALS

Student Station

watered lima bean plant
wilted lima bean plant
fine-tip forceps
prepared slide of *Nerium* (oleander)
leaf cross section

Each Group of Four Students

microscopic slides and coverglasses
dropper bottle with water

OBJECTIVES

1. To observe stomata in living bean leaves—a homeostatic mechanism for controlling transpiration and exchange of gases.
2. To compare stomata in a living, turgid, bean plant and in one that is wilted.

3. To compare stomata from mesophytic and xerophytic plants on prepared slides or from fresh materials.
4. To investigate the relationship between activity and pulse rate in humans.

Part 1
Plant Stomata: A Homeostatic Mechanism

A. *Morphology and Structure*

Remove a leaf from a well-watered bean plant. Mount a small piece of it (lower surface facing up) on a slide in a drop of water. Place a coverglass on the leaf and observe it with the low power of the microscope.

1. *Where are stomata located on the leaf?*

2. *Describe the appearance of a stoma.*

3. *Are the stomata open or closed?*

4. *What visible cellular structure do you see in guard cells that are not found in the surrounding epidermal cells?*

THE INTERACTION OF CONTROL SYSTEMS: HOMEOSTASIS 101

B. *Effects of Wilting on Stomata*

Place a plain glass slide on the stage of the microscope. Place a small bean leaf which is still attached to a well-watered plant on the slide. With the low power of the microscope, observe the condition of the stomata.

1. *Are the guard cells turgid or limp?*

2. *Are the stomata open or closed?*

Repeat this procedure with a leaf from a wilted bean plant.

1. *Are the guard cells turgid or limp?*

2. *Are the stomata open or closed?*

3. *What is the relationship between wilting and whether the stomata are open or closed?*

4. *From your observations, do you think that transpiration will increase or decrease as a plant begins to wilt?*

5. *Is wilting an example of a homeostatic mechanism? Explain.*

C. *Adaptations in Stomata of Xerophytes*

Some plants can survive where there is little available water. Certain features of the stomata show adaptations for this type of existence. Adaptations of this sort can be seen in a slide of a cross section of a *Nerium* leaf.

1. *Where are the stomata located in a* Nerium *leaf?*

2. *How does the location of the stomata aid in the conservation of water?*

3. *Note the hairs surrounding the stomata. How do they help prevent the loss of excessive amounts of water?*

Part 2
Homeostasis in Animals: Pulse Rate

Remain as still as possible for 5 min. Have someone take your pulse. (Count the beat in the wrist for 15 sec and multiply the result by 4 to obtain the pulse rate per minute.) Record the result in Table 10.1.

Go through the motions of running in place for 5 min. Take a pulse reading again and record the results in the table.

Remain as still as possible. Take pulse readings at 1-min intervals for 5 min or more if necessary until the pulse rate returns to its former rate (at rest before exercise), and record the results.

THE INTERACTION OF CONTROL SYSTEMS: HOMEOSTASIS

Table 10.1

	At Rest before Exercise	Immediately after Exercise	At Rest after Exercise				
			1 min	2 min	3 min	4 min	5 min
Pulse per minute							

1. *What produces the pulse?*

2. *Was there a substantial increase in the pulse rate after exercising?*

3. *Of what significance is the increase in pulse rate to the homeostatic control of the human body?*

4. *How long did it take for the pulse rate to return to the rate at rest before exercise?*

5. *Compare data with other members of the class. Did it take the same amount of time in each case for the pulse rate to return to its former rate?*

6. *If there are differences, what explanation can you give for the differences?*

SUMMARY

1. In this exercise examples of homeostasis in plants and animals were observed. The stoma is an example of a homeostatic mechanism in plants. The amount of water lost and the exchange of gases are controlled by the opening and closing of the guard cells. Many plants have adaptations in the stomatal apparatus which permits them to survive in very dry environments.
2. The pulse (heartbeat) contributes to steady-state control in man. The pulse rate increases with exercise and decreases when there is a cessation of physical exertion.

REFERENCES

Esau, Katherine, *Plant Anatomy*, John Wiley and Sons, New York, 1953.
Fuller, Harry J., and Oswald Tippo, *College Botany*, Henry Holt and Co., New York, 1954.
Pace, Donald M., and Benjamin W. McCashland, *College Physiology*, Thomas Crowell Co., New York, 1955.
Simpson, George Gaylord, and William S. Beck, *Life: An Introduction to Biology*, Second Edition, Harcourt, Brace and World, New York, 1965.

PROJECTS

1. Repeat Part 2 of this experiment. With a sphygmomanometer, record the blood pressure and compare the results to those obtained by taking the pulse in the wrist.
2. Obtain leaves from various xerophytes and prepare cross sections and epidermal strips. Compare their xerophytic adaptations as shown by the stomatal apparatus. If fresh materials are not available, study prepared slides.

EXERCISE 11
Communication and Behavior

MATERIALS

Each Group of Four Students

6 small aquaria or gallon bottles containing a layer of small gravel, a perch (a small stick), a 1-in. sponge cube, and a cover
3 mature male and 3 mature female *Anolis* lizards, one in each container
mealworms or flies for the lizards
cardboard separators between the *Anolis* cages
small quantity of soaked corn seeds
squares of window pane glass 6 in. X 6 in.
paper towels or blotters
masking tape
bean seedlings in small containers
4 shoeboxes or equivalent-sized cardboard boxes
scissors

Class Stock

2 large terraria, each containing 3 mature male *Anolis* and 3 mature female *Anolis* lizards. Each terrarium should contain a gravel floor, perches, and a daily water

supply for the lizards (small potted plants that can be sprinkled with water droplets are suitable)

OBJECTIVES

1. To observe social behavior and communication in the *Anolis* lizard.
2. To experiment with behavior in plants as exemplified by various tropisms.

Part 1
Social Behavior and Communication in the *Anolis* Lizard

Anolis lizards are common inhabitants of the southeastern United States. They are usually available from biological supply houses and can be kept alive in a terrarium for considerable periods of time. Mature males can be distinguished by their larger throat fan. *Anolis* exhibits simple but easily observed behavior patterns that illustrate some of the concepts of animal communication and interaction.

An inexpensive cage consists of a wide-mouth gallon jar covered with screen. The jar is equipped with a perch (piece of tree limb) and a layer of fine gravel on the bottom. Each *Anolis* is fed several times a week on mealworms, houseflies, or other small insects. For water, a moist sponge cube is suspended within reach of the lizard. The sponge must be dampened daily so that water droplets are available to the *Anolis*. In nature the lizard drinks water droplets from leaf surfaces.

A week prior to this lab period, three male and three female *Anolis* lizards should be placed together in a large terrarium. They are color coded with small dabs of dye so that each can be identified. Three or four students should work together in observing and recording the behavioral activities of these lizards throughout the exercise. The group should first watch undisturbed lizards for a short period of time to become familiar with their normal or usual behavior patterns. Especially look for color changes, enlargement of the throat fan, elevation of the nuchal (neck) crest near the back of the head, opening of the mouth, head bobbing, and other body movements that seem to be repeated as a normal part of the lizard's behavior and interaction with other cage members. Record your observations in Table 11.1.

In order to see some of the typical *Anolis* reactions, it is necessary to manipulate the lizards. This is especially true in relation to dominance and territoriality behavior. Do not repeat the experiments unnecessarily; the lizards will cease to respond.

Gently push one of the males closer to another male. Repeat this procedure with other males.

1. *Did you observe a dominance reaction? If so, describe how each lizard reacted to the other one.*

2. *What seems to determine which lizard will be dominant in each case?*

3. *What sort of communication is involved here?*

4. *In these reactions you were disturbing territorial relations which the lizards established in the terrarium. How does a lizard communicate to another male that it is trespassing?*

5. *From your observations, suggest a technique for mapping a lizard's territory.*

6. *Gently push one female close to another female and describe the results.*

7. *Try the other females. Do any of them react like the males did?*

108 EXPERIMENTS IN FUNDAMENTAL CONCEPTS OF BIOLOGY

8. *What do you conclude about territoriality among female* Anolis?

9. *Is there any observable "communication" between the two?*

10. *Introduce a new lizard into the terrarium and describe the results.*

	Table 11.1					
	Male 1	Male 2	Male 3	Female 1	Female 2	Female 3
Marking code						
Color changes						
Enlargement of throat fan						
Elevation of nuchal crest						
Opening of the mouth						
Head bobbing						
Other body movements						
Peck order dominance						
Territoriality						

11. *Does it make any difference whether the new member is a male or female? Why?*

Permit the lizards to recover from your manipulations (15 to 30 min). Then place a container of food in the center of the terrarium. Observe the feeding relations among the lizards.

1. *Can you observe a peck order among the lizards in terms of which ones feed first? If so, describe it and the dominance hierarchy that seems to apply.*

2. *How is this different from the dominance relationship in territorial behavior?*

3. *In all the reactions observed so far, what is the major, perhaps the only, form of communication?*

4. *Could the lizards be communicating by some means, chemical, for example, that we have not detected? If so, what means?*

5. *How could you go about discovering a mode of communication that is not immediately evident or easily observed?*

110 EXPERIMENTS IN FUNDAMENTAL CONCEPTS OF BIOLOGY

The dominance relations and reactions can be explored further using lizards in individual jars. These are placed side by side, separated by a piece of cardboard. When the cardboard is removed, the lizards may exhibit dominance behavior. Using three males and three females in separate containers, test their reactions to one another in this fashion. Attempt to establish an order of dominance by this technique.

The behavior you have been observing is controlled by hormones. Lizards that react poorly, or not at all, can be stimulated into activity by the administration of sex hormones. Several techniques for doing this are described in the BSCS Laboratory Block termed *Animal Behavior*.

Part 2
Tropisms in Plants

In Exercise 8 you observed two aspects of chemical control in plants, the site and activity of auxins. Here we want to explore some of the activities of auxins in response to environmental stimuli. A tropism is the growth movement a plant makes toward or away from a stimulus. This differential growth is mediated by auxins.

A. Response to Water (Hydrotropism)

The following procedure can be performed with either radish or corn seeds. Cut some paper towels or blotters to fit the squares of window glass provided at your lab station. Fold one of the paper towels in half, wet it, and position it on a glass plate. Place soaked seeds along the edge of the paper, as shown in Fig. 11.1. Fold a dry towel and place it along the other side of the seeds. Use the other square of glass for a cover and tape all four edges of the two plates. Lay the plates flat and observe for several days.

1. *Do the roots show a positive or a negative hydrotropism?*

2. *How do you know that this was not a gravity effect?*

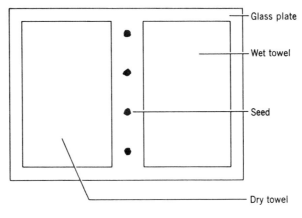

Figure 11.1

3. *How is this tropism beneficial to the plant?*

4. *Do the seed shoots show a reaction to water in this experiment?*

B. *Response to Gravity (Geotropism)*

Cut several paper towels to fit a glass plate. Wet them and place on the plate. Position four soaked corn seeds so that each points in a different direction, as shown in Fig. 11.2.

Cover with another glass plate and seal the edges with tape as before. Stand the plates *on edge* and observe for several days.

1. *How do the shoots react?*

2. *Is this positive or negative geotropism?*

3. *Can you determine how the roots react?*

4. *After the shoots establish a definite geotropism, stand the plates on a different edge. How do the shoots respond to this change in position?*

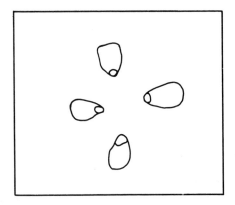

Figure 11.2

Perhaps these reactions are typical only of germinating seeds. To test this, obtain some bean seedlings growing in small containers and lay them on one side. Observe the results after a few days.

1. *How did the bean seedlings react to gravity?*

2. *How is geotropism beneficial to a plant?*

C. *A Conflict Situation: Hydrotropism Versus Geotropism*

Arrange a set of plates and towels as in part A. Place soaked corn seeds at the junction between a wet and a dry blotter. Make sure that all the seeds are oriented with their smaller ends toward the dry blotter. Seal the plates and stand them on edge so that the damp side is up, as in Fig. 11.3. Observe for several days, and answer these questions:

1. *To which stimulus did the roots respond?*

2. *To which stimulus did the shoots respond?*

3. *How are these reactions advantageous to a plant?*

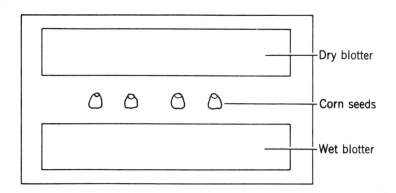

Figure 11.3

114   EXPERIMENTS IN FUNDAMENTAL CONCEPTS OF BIOLOGY

D. Response to Light (Phototropism)

This set of experiments demonstrates the response of plant shoots to light. Obviously, this tropism is vital to a plant's survival.

Obtain four bean seedlings in separate containers and four "growth chamber" boxes. Three boxes should be arranged as shown in Fig. 11.4. The fourth box is to be an experiment devised by you.

Set the boxes near a good source of light. Open only once per day to water the plant and to observe it. At the end of 1 week indicate the results by diagramming the plant's pattern of growth in each experiment in Fig. 11.4. Account for the results you obtained in each experiment.

Experiment 1.

Experiment 2.

Experiment 3.

Experiment 4.

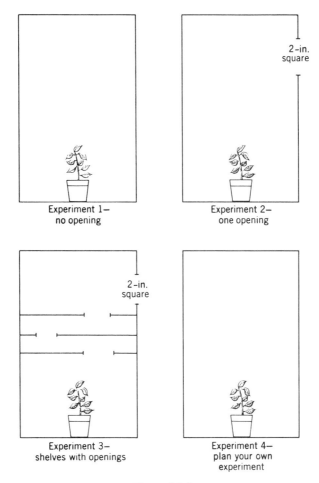

Figure 11.4

1. *Would you say that stems give a "strong" response to light? Why?*

2. *Why is this tropism obviously vital to plants?*

3. *Would you expect roots to show a reaction to light? Why?*

4. *Why was the box in Experiment 1 set up with no openings?*

SUMMARY

1. *Anolis* lizards illustrate animal communication based on visual signals. These signals are of various kinds, each conveying a specific bit of information concerning the behavior of *Anolis*. Dominance and territoriality behavior are controlled by hormones in *Anolis* and function only in males of this species.
2. Plants exhibit a variety of adaptive behavior mechanisms, mostly controlled by hormones produced in root and stem tips. These involve environmental factors basic to a plant's survival such as light, water, and gravity.

REFERENCES

Follansbee, Harper, *A Laboratory Block. Animal Behavior*, D. C. Heath and Company, Boston, Mass., 1965.

Lawson, Chester A., and Richard E. Paulson, *Laboratory and Field Studies in Biology*, Holt, Rinehart and Winston, New York, 1960.

PROJECTS

1. Capture lizards native to your area and repeat the *Anolis* experiments with them. Try to observe and describe their signals and forms of communication.
2. Aquarium fishes often show clear territory and dominance behavior. Capture some native fishes and observe their behavioral peculiarities. Some small sunfishes, for example, establish territories in an aquarium and defend them vigorously with interesting displays and aggressive behavior.
3. Design unusual environmental variations or conflict situations and observe a plant's response to them. Here are a few examples: (*a*) Place a plant in a miniature merry-go-round or ferris wheel and note the effects of gravitational stimuli. (*b*) Place a light under a plant and observe its growth. (*c*) Design an experiment to show whether roots respond more strongly to gravity or to a nutrient chemical solution.

EXERCISE 12
Asexual Reproduction

MATERIALS

Student Station

scalpel
acetocarmine stain in dropper bottle
rusty dissecting needle (If iron acetocarmine stain is available, rusty needle needed.)
microscope slides and coverglasses
paper towels
Allium (onion) root tip mitosis, prepared slide
whitefish blastula mitosis, prepared slide

Each Group of Four Students

onion root tips (See Fig. 12.1)
alcohol lamp
small flat or clay pot filled with equal parts of sand and peat
several Irish potatoes
3 planaria

4 Syracuse watch glasses
pond water

OBJECTIVES

1. To observe mitosis in onion root tip and whitefish blastula.
2. To investigate asexual propagation of potatoes.
3. To observe regeneration in planaria.

Part 1
Mitosis: Cellular Reproduction

Mitosis is a process by which the nucleus divides into two nuclei, each of which contains identical genetic material. Usually the cell divides also. See *Fundamental Concepts of Biology,* pages 168-170. Although mitosis is a continuous process, it is convenient to speak of distinct phases—prophase, metaphase, anaphase, and telophase. Interphase refers to the nondividing nucleus after or before mitosis. Each phase has distinguishing characteristics. Bear in mind as you study the slides that in life one phase flows into the next with no distinct breaks between them.

A. Onion Root Tip

Five days in advance of this experiment, place onions in small glass containers with water. See Fig. 12.1.

Remove an onion from the container and cut a tip 5 mm long from a root. Place it on a microscope slide and cut into small pieces. Add a few drops of acetocarmine stain. Place a rusty dissecting needle in the solution to allow iron to get into the stain, or use an iron acetocarmine stain. Add a coverglass and place a small piece of paper towel over the coverglass. Gently mash the coverglass with the thumb, being careful not to break it. Lift the coverglass and add a few more drops of acetocarmine. Replace the coverglass and blot any excess dye with a paper towel. Pass the slide quickly over the flame of an alcohol lamp two or three times until the stain darkens. Do not overheat. Locate mitotic figures with the low power of the microscope. Turn to high power for details.

1. *What do the chromosomes look like?*

ASEXUAL REPRODUCTION 119

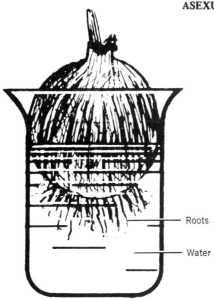

Figure 12.1 Onion suspended in water.

2. *Locate chromosomes in prophase, metaphase, anaphase, and telophase. Describe the arrangement of the chromosomes in each stage.*

Prophase.

Metaphase.

Anaphase.

Telophase.

3. *Why did the directions specify using* root *tips?*

Study prepared slides of onion root tip mitosis. These slides will show the mitotic figures *within* the cells.
Locate a prophase cell.

1. *How are the chromosomes arranged?*

2. *What happens to the nuclear membrane?*

3. *Do you see spindle fibers?*

4. *Are there centrioles?*

Locate a metaphase cell.

1. *How are the chromosomes arranged?*

2. *What is the name of the portion of the cell where the chromosomes are located?*

3. *How can you determine the poles of the cell?*

Locate an anaphase cell.

1. *Describe the position of the chromosomes.*

2. *Even though it appears that the original number of chromosomes has been reduced by one-half, this is not true. Each daughter nucleus inherits the same number of chromosomes originally present in the parent nucleus. How do you explain this phenomenon?*

Locate a telophase cell.

1. *What criteria did you use to recognize the telophase cell?*

Locate an interphase cell.

1. *Are the chromosomes visible?*

2. *Describe the appearance of the cell.*

B. *Whitefish Blastula*

Study a prepared slide labeled whitefish blastula and locate the four stages of mitosis. Note that centrioles, absent in onion cells, are present in whitefish cells.

1. *What is the position of the centrioles in prophase?*

2. *Where are the centrioles located in metaphase?*

3. *How is cytoplasmic division (cytokinesis) accomplished in whitefish?*

4. *How does this differ from cytokinesis in onion?*

Part 2
Asexual Propagation of Potatoes

Obtain an Irish potato and cut it into four pieces as follows:

1. Cut out an eye. Remove all starchy material from it.
2. Cut a 1-in. piece of potato containing one eye.
3. Cut a 1-in. piece of potato. Remove eyes.
4. Cut a 2-in. piece of potato so that it includes at least two eyes.

Plant the four pieces of the potato in a flat or a clay pot containing equal parts of sand and peat. Keep moist at all times. At the end of 1 week observe the results and record your observations in Table 12.1.

Table 12.1

	Observations
Potato eye alone	
Potato with one eye	
Potato with no eyes	
Potato with two or more eyes	

1. *In which treatments did growth occur?*

2. *Why did growth not occur in all treatments?*

3. *What plant organ is the Irish potato?*

124   EXPERIMENTS IN FUNDAMENTAL CONCEPTS OF BIOLOGY

4. *What anatomical part is the eye of the potato?*

5. *What is the function of the starchy material?*

6. *If you were going to grow potatoes commercially by asexual propagation, which method of the four tried in this experiment would you use? Discuss the reason for your choice.*

7. *What advantage is there to propagating plants asexually?*

Part 3
Regeneration in Planaria

Obtain three planaria. Treat in the following ways:

1. Cut one worm longitudinally with a scalpel (See Fig. 12.2).
2. Cut the second worm across the center, separating the anterior end from the posterior end (see Fig. 12.3).
3. Cut the third worm into many small pieces.

Place the treated worms into Syracuse watch glasses containing clear pond water. Label the containers, stack the dishes, and place an empty one on top to discourage evaporation of water. At the end of 1 week observe and record the results in Table 12.2.

ASEXUAL REPRODUCTION 125

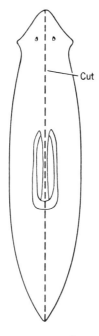

Figure 12.2

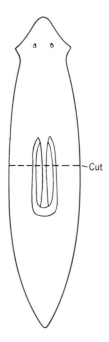

Figure 12.3

Table 12.2
Observations

Treatment 1—
longitudinal cut

Treatment 2—
cross-section cut

Treatment 3—
pieces

1. *Did regeneration occur in all treatments?*

2. *What cellular event is involved in regeneration?*

3. *To what extent does regeneration occur in a higher animal such as humans?*

4. *Of what adaptive value is regeneration?*

SUMMARY

Mitosis is the process by which cells are reproduced asexually. The daughter cells are identical to the parents. Most plants may be propagated asexually by taking pieces from the parental tissue. Regeneration in planaria and many other animals will occur when parts are lost. Asexual propagation of potatoes and regeneration in planaria are dependent upon the process of mitosis.

REFERENCES

Johansen, Donald Alexander, *Plant Micro-technique,* McGraw-Hill, New York, 1940.

Turtox Service Leaflet No. 16. *The Culture of Planaria and Its Use In Regeneration Experiments,* CCM: General Biological Inc., 1959.

PROJECTS

1. Investigate the effects of colchicine on mitosis in onion root tip. See article by Norman J. Gillette in *Turtox News* [43(3), 90 (March, 1965)] for directions.
2. Propagate various plants by soft-wood cuttings and leaf cuttings. Compare the effectiveness of the methods for the plants involved.
3. Try additional cuts on planaria. Interesting results can be obtained, such as two-headed planaria. See Turtox Service Leaflet No. 16 for information about how to make the cuts.

EXERCISE 13
Sexual Reproduction

MATERIALS

Each Group of Four Students

living or preserved specimens of *Mnium, Polytrichum,* or other appropriate mosses
microscope slides and coverglasses
dissecting needles
methylene blue in dropper bottle
prepared slides of moss archegonia, antheridia, and protonema

Class Stock

prepared slides of meiosis in plant and in animal tissues
suspension of frog sperm

OBJECTIVES

1. To observe some of the major stages of meiosis in animal and plant tissues.
2. To examine the stages in the life cycle of a moss.
3. To observe living sperm cells.

128   EXPERIMENTS IN FUNDAMENTAL CONCEPTS OF BIOLOGY

Part 1
Meiosis

In the previous exercise, mitosis was observed in plant and animal tissues in considerable detail. Stages of meiosis are not as easily obtained or observed, hence this section of Exercise 13 will probably be presented as demonstration material. The instructor has set up one or more slides showing stages or features of meiosis he wishes to emphasize. Look at the demonstrations and summarize what you see with a short description or some simple diagrams.

Meiosis in plant tissue:

Meiosis in animal tissue:

Part 2
Life Cycle of a Moss Plant

The life cycle of a moss such as *Mnium* or *Polytrichum* provides convenient study material for examining several basic aspects of plant reproduction. Gametophyte and sporophyte generations are easily studied, as are their specialized reproductive bodies. In addition, suitable local species are nearly always available.

SEXUAL REPRODUCTION 129

You will find the illustration in *Fundamental Concepts of Biology* (page 221) of the life cycle of *Mnium* helpful as you proceed with the following directions.

Obtain a mature male moss plant and a mature female moss plant and examine them closely.

1. *How can you distinguish between the two?*

2. *To which generation do these two plants belong?*

3. *Would you term this generation sexual or asexual? Why?*

Look at prepared slides showing male reproductive bodies (antheridia) and female reproductive bodies (archegonia). These slides are made by sectioning the tip of the plant and staining the tissues to emphasize structural details.

1. *Does meiosis occur within archegonia and antheridia? Why?*

2. *How does the production of sperm per antheridium compare to the number of eggs per archegonium?*

3. *What is the adaptive value of this?*

Hold the tip of the mature female moss plant in a few drops of water on a microscope slide. Gently crush it and tease out its tissues with a dissecting needle. Cover the teased tissue and examine under a microscope. Describe the appearance of the archegonia.

Perform the same manipulation with the end of the male plant. Describe the appearance of the antheridia.

1. *What adaptive relationship, if any, exists between the habitat of most mosses and the means by which fertilization is brought about?*

2. *After the egg is fertilized in the archegonium, what structure does the zygote form?*

Obtain a mature sporophyte specimen and examine it.

1. *On what does the sporophyte depend for nourishment and support?*

2. *Are the sporophyte cells haploid or diploid? Why?*

Crush a spore case (sporangium) in a drop of water and observe the spores under the microscope. Normally they are released into the air.

1. *How are they adapted to facilitate dispersal?*

2. *The formation of spores in the sporangium is the consequence of what basic reproductive event?*

Obtain a prepared slide of the tiny structure into which a spore grows (protonema).

1. *What sort of habitat would this structure probably require?*

2. *What will it eventually form?*

3. *The presence of sporophytes and gametophytes in the life cycle is termed* alternation of generations. *Explain why.*

Part 3
Living Sperm Cells

To avoid sacrificing an excessive number of frogs, the laboratory instructor has prepared a suspension of frog sperm. To prepare it, testes were removed from a frog and cut into small pieces in a quantity of pond water. The pieces were mashed to free the sperm. In 10 to 15 min the sperm should be active.

Place a drop of the suspension on a microscope slide under a cover slip. Examine carefully and describe the activity of the sperm cells.

1. *Is their motion erratic or do they move in straight pathways?*

2. *Is the organ of locomotion (tail) clearly visible? Why?*

While the sperm are still active on your slide, add a drop of methylene blue stain.

1. *Which part of the sperm cell now becomes more easily seen?*

2. *What effect does the dye have on the sperm?*

SUMMARY

1. Meiosis occurs in plants and animals in tissues specialized for reproduction. Meiotic stages observed on microscope slides resemble many of the mitotic stages observed in Exercise 12. Meiosis is far more complex, as a careful study of appropriate slides would show. For example, it is possible to see the paired homologous chromosomes, the arrangement of chromosomes on the Metaphase I plate, and the four daughter cells resulting from Telphase II. These stages are not easily found on most prepared slides and thus are often presented as demonstration material.
2. Mosses provide convenient material for observing alternation of generations in plants. In mosses these generations are about equal in size. Moisture is necessary

for sperm to reach archegonia. Dispersal of the plant into new environments is aided by the release of tiny airborne spores.
3. Frog sperm cells are motile and microscopic in size. Motility occurs from the action of a specialized area of cytoplasm in the tail.

REFERENCES

Bold, Harold C., *Morphology of Plants,* Harper and Bros., New York, 1957.

Morholt, Evelyn, Paul F. Brandwein, and Alexander Joseph, *A Sourcebook for the Biological Sciences*, Harcourt, Brace and Company, New York, 1958.

PROJECTS

1. Using unfertilized frog eggs, attempt to stimulate their development (cleavage) by artificial means—pricking with a needle, agitation, abrupt temperature changes, to suggest a few. If successful, attempt to raise some haploid frogs.
2. Collect local species of mosses. If the sporophyte stage is available, attempt to culture some spores using media described under References.
3. Examine sperm cells from other organisms (insects are a good source) and compare their general morphology and motility.

EXERCISE 14
Development

MATERIALS

Student Station

compound microscope
stereomicroscope
starfish *(Asterias)* prepared slides as follows:
    blastula whole mount
    gastrula whole mount
    bipinnaria whole mount (optional)
dropper

Each Group of Four Students

male frog
culture dishes
1 10-ml graduated cylinder
microscope slides and coverglasses
2 Syracuse watch glasses
glass rod
bean seedling in clay pot

136   EXPERIMENTS IN FUNDAMENTAL CONCEPTS OF BIOLOGY

Class Stock

3 female frogs (treated with pituitary extract to induce ovulation)
pond water known to support frog life
sea urchins containing mature eggs and sperm
artificial sea water
bean seeds (soaked in water for 24 hours)
germinating bean seeds (grown in petri dishes between wet filter paper)

OBJECTIVES

1. To examine early embryonic stages in the development of starfish.
2. To observe early cleavage in living frog eggs.
3. To observe early cleavage in living sea urchin eggs.
4. To examine selected stages in the development of a bean plant.

Part 1
Starfish Development (*Asterias*)

Although there is great diversity of forms in the animal kingdom, most animals follow a basic similar plan in embryonic development, at least in the early stages of development. The fertilized egg, a single cell, undergoes a series of mitotic divisions (cleavage) to form an embryonic mass called a blastula, later a gastrula. Finally, by differentiation and further divisions and growth, an adult is produced. Some organisms go through a larval stage before changing into the adult form. This is true of the starfish.

A. Asterias

Look at a slide labeled *Asterias* (starfish) early and late cleavage whole mount. Locate and diagram an unfertilized egg, a fertilized egg, a two-cell stage, and a four-cell stage. Observe an eight-cell stage, a sixteen-cell stage, and a morula. (See *Fundamental Concepts of Biology,* page 258.)

1. *What is the difference in appearance between an unfertized egg and a fertilized egg?*

2. *Each cell in the embryo during the cleavage phase of development is called a blastomere. Are the blastomeres in the four-cell stage the same size as the blastomeres in the two-cell stage?*

3. *How does the size of the blastomeres in the sixteen-cell stage compare with those in the eight-cell stage?*

4. *Explain your answers to questions 2 and 3 above.*

5. *Compare the overall sizes of the two-cell, the four-cell, the eight-cell, and the sixteen-cell stages and the morula. Are they of different sizes or are they all alike? In view of your answer to question 4, did you expect this to be true? Why?*

B. Starfish Blastula

Examine a slide of starfish blastula whole mount.

1. *Is the blastula larger, smaller, or the same size as the early cleavage stages?*

2. *What shape does the blastula assume?*

3. *Notice that the cells are arranged to form a hollow space at the center. What is this cavity called?*

C. *Starfish Gastrula*

Examine a slide of starfish gastrula whole mount. The gastrula is formed from the blastula by a process called *gastrulation*. The particular kind of gastrulation that occurs in starfish is called invagination. This process is accomplished by the movement of cells on one part of the blastula into the cavity. This has been likened to a partially deflated basketball with one side punched in. Diagram a gastrula. Locate and label the following parts of your diagram: blastocoel, archenteron, ectoderm, endoderm. (Refer to *Fundamental Concepts of Biology,* page 258.)

1. *What will eventually happen to the blastocoel?*

2. *What is the archenteron?*

3. *Look at a number of gastrulae. Are they all the same size?*

The gastrula develops into a swimming dipleurula larva and later into the adult starfish. If a slide of a larva is available, examine it.

Part 2*
Early Cleavage in Frog Development

(Work in groups of four.)

A. The Male Frog

Pith a male frog and dissect out the testes. The testes can be recognized by their oval shape and yellow color and will be found on the dorsal surface of the body cavity near the kidneys. Macerate the testes in a culture dish containing 11 ml of clean pond water. After 5 min place a drop of sperm suspension on a microscope slide and cover with a coverglass. Observe the slide with the low power and then the high power of the microscope. If the sperm cells are motile, proceed with Part B. If not, wait a few minutes and make another slide.

B. The Female Frog

Obtain a female frog which has been treated to induce ovulation. Hold the frog's body with one hand and its legs with the other. Press its abdomen firmly with the thumb and the eggs should be released. Milk the eggs into a culture dish which contains sperm suspension. Pipette the sperm suspension over the eggs with a dropper. Place the culture dish under the stereomicroscope and observe. The eggs are held together in a mass by a jellylike substance. Of what adaptive value is this substance if the eggs had been deposited under normal conditions in a pond?

---

*It is suggested that either part 2 or Part 3 be carried out, depending on the organism available.

C. Addition of Pond Water

Add enough pond water to cover the eggs, and let stand for 15 min. Then change the water. Try to rotate an egg. Notice that one side is light colored (contains yolk), and the other side is a dark color. When the eggs are deposited they are oriented in various ways. However, if fertilization has been accomplished, in about 1 hr the eggs rotate so that the dark side is oriented upward. Of what possible adaptive value is this in nature? *(Hint:* What do you know about the ability of various colors to absorb heat?)

D. Observation

After the eggs have rotated, separate them into small numbers, about 20 per batch, and place into two separate dishes containing pond water. Observe the eggs with the stereomicroscope from time to time during the laboratory period.

1. *How long did it take, from the time the eggs were fertilized, for the first cleavage to occur?*

2. *Note that the second cleavage begins before the first one has completely divided the fertilized egg into two blastomeres. Why do you suppose the process slows down in the lower half?*

The frog embryonic development proceeds from the early cleavage stages to a blastula, then a gastrula, and finally to a larva which hatches from the capsule. It later undergoes metamorphosis into an adult. Although there are differences, there is a basic similarity between the development of a frog and the development of a starfish. Can you think of one or more reasons for the presence of a larval stage in the development of some organisms?

Part 3
Early Sea Urchin Development*

(Work in Groups of Four.)

A male and a female sea urchin have been provided for each group. The instructor will demonstrate the procedure for obtaining the eggs and sperm. Allow the sperm to drop into a clean, dry Syracuse watch glass. The eggs should be deposited into a small culture dish containing 30 ml of artificial sea water adjusted to the specific gravity similar to the sea water where the sea urchins were collected. Sea water should not be used unless it is first processed to remove contaminants. Rotate the culture dish gently to wash the eggs. They should collect at the center of the dish. Carefully pour off the water and add fresh sea water. Repeat the operation twice.

Touch the sperm suspension with a glass rod and transfer a very small amount to a Syracuse watch glass containing sea water. Then drop a few eggs into this watch glass. Place the watch glass on a compound microscope and focus on the eggs as quickly as possible. Be careful not to allow the objective to touch the water. Use the low power only; do not attempt to use high power. After fertilization a membrane appears around the egg. Locate eggs that have been fertilized. Many sperm can be seen swimming about. Some will be seen imbedded in the fertilization membrane. Note the difference in size between the sperm and the eggs. What adaptive significances can you suggest for the differences in size and numbers between the eggs and the sperm?

The development of the sea urchin is similar to the development of the starfish. As you observe cleavage, review Part 1 and compare the diagrams you made to what you see in the living material.

---

*Note to instructor. The authors have found that *Arbacia* is satisfactory during the summer months. Eggs and sperm will be shed if the animal is held so that the oral opening is up and a mild electrical stimulus is applied to the oral opening. The eggs will be shed through the aboral sinus. With this sea urchin the sexes can be determined without sacrificing the animal. The eggs will be red and granular in appearance as they are shed. The sperm will be white and not at all granular in appearance. *Lytechinus* is satisfactory only during the colder months. To obtain the eggs and sperm, cut out the Aristotle's lantern and add a few drops of 0.5 M KCl to the cavity. The sex can be determined as before by the appearance of the eggs and sperm or by looking into the cavity of the sea urchin. The differences in appearance will be apparent in the gonads.

Part 4
Bean Seedlings

The bean plant is a dicot. This is the shortened form of the word dicotyledon, which means two seed leaves. The function of cotyledons is to store food the seedling uses until the mature leaves are able to carry out photosynthetic activities.

Examine a bean seed which has been soaked in water for 24 hours. Remove the seed coat carefully and separate the cotyledons from one another. The embryo will be found adhering to one of the cotyledons. Look at the embryo with the aid of a hand lens or the stereomicroscope. Identify the epicotyl and the hypocotyl. (See *Fundamental Concepts of Biology*, page 261.)

Secure a germinating seed from the petri dish and a young seedling from the flat. Carefully wash the soil away from the roots. Identify the cotyledons, the hypocotyl, and the epicotyl on each.

Observe the older bean plant in the clay pot. Identify cotyledons, the hypocotyl, and the epicotyl.

1. *In what form is food stored in the cotyledons? How could you test the validity of your answer? (See Exercise 2.)*

2. *The portions of the epicotyl that will develop into the first foliage leaves are called the plumules. Why are not the plumules green?*

3. *Into what structure will the hypocotyl develop?*

4. *What happens to the cotyledons as the seedling grows? Why?*

5. *What portion of the embryo makes its appearance first after germination? Can you suggest the significance of this adaptation?*

6. *If you tied a string around the bean stem 1 in. above the surface of the soil, at what distance from the soil would the string be at the end of one week? Explain.*

SUMMARY

1. Although considerable diversification of adult forms exists in the animal kingdom, there are basic similarities among many of them insofar as embryonic development is concerned. This is well illustrated by the starfish, the sea urchin, and the frog. All three start development as a fertilized egg which undergoes a series of mitotic divisions to form a blastula, gastrula, larva, and finally an adult.
2. Bean plant development illustrates the basic scheme common to seed plants. The cotyledons provide energy for the developing seedling until leaves can take over that function.

REFERENCES

Humphrey, Donald G., Henry Van Dyke, and David L. Willis, *Life in the Laboratory,* Harcourt, Brace and World, New York, 1965.

Machlis, Leonard, and John G. Torrey, *Plants in Action,* W. H. Freeman and Co., San Francisco, 1959.

Patten, Bradley, *Foundations of Embryology,* McGraw-Hill, New York, 1958.

Rugh, Robert, *The Frog,* The Blakiston Co., Philadelphia, Pa., 1951.

Wodsedalek, J. E., *General Zoology Laboratory Guide,* Fourth Edition, Wm. C. Brown Co., Dubuque, Iowa, 1955.

PROJECTS

1. Save some of the frog embryos from this exercise and allow them to grow into tadpoles. This may be done by transferring the embryos to an aquarium containing pond water. Feed the larvae boiled lettuce. Remove excess lettuce so that it will not foul the water. Change the water occasionally. Depending upon temperature, tadpoles should be obtained in about 10 days.
2. Try to obtain pluteus larvae from the sea urchin embryos by allowing the embryos to develop in covered Syracuse watch glasses. Be sure the sea water does not evaporate. Add more if necessary. No food is necessary. Three to four days are required to obtain larvae.
3. Investigate the growth patterns of the bean stem, roots, and leaves. Grid the immature leaf with India ink at intervals of 2 mm (see Fig. 14.1) and measure the distance between the lines at the end of 1 week. Note whether growth is evenly distributed. Mark the hypocotyl of a bean seedling with India ink at intervals of 1 mm for a length of 10 mm from the tip (see Fig. 14.2). Measure the distance between the lines after 48 hr. Observe whether the growth was evenly distributed.

Figure 14.1 Leaf with India Ink grids.

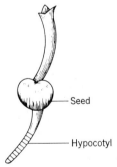

Figure 14.2 Bean seedling.

EXERCISE 15
Structures Associated With Heredity

MATERIALS

Student Station

flower for dissection (living or preserved)

Each Group of Four Students

male frog, pithed or preserved
female frog, pithed or preserved
2 dissecting pans
2 dissecting needles
supply of pins
2 scalpels

Class Stock

model of flower

146   EXPERIMENTS IN FUNDAMENTAL CONCEPTS OF BIOLOGY

OBJECTIVES

1. To dissect and observe the reproductive structures associated with the urogenital system of the frog.
2. To observe the reproductive structures of the flower—location of chromosomes involved in transmission of genetic traits.

Part 1
The Reproductive System of the Frog

Work in groups of four. Each group will be given a male and a female frog. Two members of the group will dissect the female frog and the other two members of the group will dissect the male frog. After completing the dissection, exchange frogs, so that every member of each group will have an opportunity to study both sexes.

Place a frog, ventral side up, in a dissecting pan. Make incisions as indicated by the dashed lines in Fig. 15.1. By pinning the flaps to the dissecting pan, the observations of the internal organs will be facilitated.

A. The Female Frog

Locate the *ovaries*, which are held in place by a mesentery (*mesovarium*). The size of the ovaries varies according to the time of the year. Identify the coiled oviducts and the opening to each oviduct, which is called the *ostium*. During the breeding season, eggs are released from the ovaries and are directed to the *ostia* by the movement of cilia. The eggs move into the oviducts where they are held temporarily in the lower portion, and then are released into the cloaca and out of the body through the cloacal opening. Fertilization takes place outside the body. Cut open the cloaca and locate the openings to the oviducts.

1. *You observed in Exercise 14 that frog eggs are covered with a gelatinous substance. Where is this substance secreted?*

2. *How could you test the validity of your answer to question 1 by observation of the dissected frog?*

STRUCTURES ASSOCIATED WITH HEREDITY   147

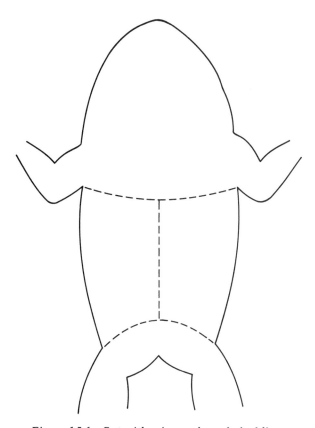

Figure 15.1 Cut with scissors along dashed lines.

3. *What is the significance of the large number of eggs which mature during one breeding season?*

4. *Two tubes in addition to the oviducts lead to the cloaca. What are they?*

5. *Locate the fat bodies near the ovaries and kidneys. What is their function?*

B. *The Male Frog*

Locate the two testes, which are held in place by the mesentery, called the *mesorchium.* The *vasa efferentia* are ducts which lead from the testes to the kidneys. Sperm pass through these ducts into the kidneys. The sperm cells are carried through the ureters to the cloaca and reach the outside through the cloacal opening. Locate the *vasa efferentia,* the *kidneys,* and the *ureters.*

1. *Why are the ureters in the male frog often called urogenital ducts?*

2. *Why is a penis not necessary to accomplish fertilization in frogs?*

Part 2
Reproduction in Flowering Plants

The chromosomes, which produce the next generation of plants, are found in the ovules and the pollen. Although there are numerous kinds of flowers, all of them exhibit a basic similarity in floral parts specialized for reproduction. Accessory structures such as petals and sepals may be present or absent. The pistil and stamens are often located in the same flower or they may be located in separate flowers on the same plant or on different plants. Regardless of the variations, it is useful to identify the reproductive structures and accessory parts if present on one type of flower and apply this information to flowers in general.

Examine a flower, living or preserved, and compare it to the flower model. (See Fig. 15.2.) Identify the following structures. The outermost whorl of structures, called the *sepals,* is usually green, sometimes leaf-like. The sepals cover the flower bud before the flower opens. Sepals collectively are called the *calyx,* and in lilies they are usually the same color as the next whorl or *petals.* The petals are often brightly colored and conspicuous. Collectively the petals are the *corolla*. The corolla and the calyx together are known as the *perianth*. The next set of structures toward the middle consists of *stamens*. Each stamen is made up of an anther, which produces pollen, and an elongated filament. Remove an anther and cut a cross section. Observe the pollen within the *pollen sacs*. The central structure in the flower is the *pistil*. It has three parts: uppermost is the *stigma*, to which pollen adhere; next is the *style*, or central portion; and the *ovulary* is at the bottom. Within the ovulary will be found *ovules*. The expanded basal portion of the flower is known as the *receptacle*, and the stalk below the receptacle is called the *peduncle*.

The general steps that result in fertilization are as follows: Pollen produced in the anther sacs of the stamen are transferred to the stigma portion of the pistil. This process is pollination. Pollen grains adhere to the stigma because the stigma produces a sticky substance. The pollen germinate and pollen tubes grow downward through the style. Each pollen tube has a tube nucleus, which regulates the growth of the tube, and a generative nucleus, which divides by mitosis to form two sperm nuclei. The pollen tube makes its way to the ovulary portion of the pistil and enters the ovule through the micropyle (small pore).

Within the ovule are eight nuclei, three at the micropylar end, three at the opposite end, and two in the central part of the sac. One of the nuclei at the micropylar end is the egg. The pollen tube entering the embryo sac of the ovule discharges the two sperm into the sac. One sperm fuses with the egg. This fertilization produces a zygote, which develops into the embryo of the seed. The other sperm fuses with the two nuclei at the center known as *polar nuclei* to form the *endosperm nucleus*. After a series of mitotic divisions, the endosperm nucleus gives rise to cells that contain *endosperm* (food for the embryo). The other five nuclei degenerate. The outer covering of the ovule becomes the seed coat. The ovulary matures into a fruit, which in time releases the seeds to complete the life cycle. (See Exercise 14.)

1. *Of what adaptive value are showy petals in flowers?*

2. *List ways in which pollen reach the stigma.*

3. *Many flowering plants have inconspicuous flowers in which petals and sepals are not present. These plants are often overlooked as flowering plants by persons with little knowledge of botany. Can you attribute any significance to the lack of conspicuous petals in plants of this type?*

4. *In some plants the embryo sac and the anther sacs of the same flower mature at different times. Of what value is this?*

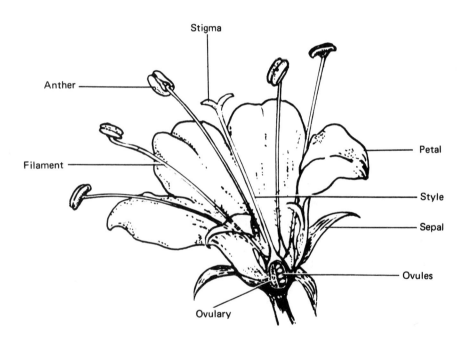

Figure 15.2 Structures of a flower.

5. *How can you account for the occurrence of an occasional undersized pea in a pod? (Recall that mature ovularies are fruits and mature ovules are seeds.)*

SUMMARY

1. The ovaries and the testes of the frog are the structures in which the gametes involved in sexual reproduction are produced. The gametes contain the chromosomes and genes responsible for transmission of hereditary traits.
2. The reproductive structures of the flower are the pistil and the stamen. The embryo sac of each ovule and the pollen chambers of anthers contain reproductive cells involved in sexual reproduction. An egg nucleus is produced in the ovule and sperm nuclei are produced in the pollen tube.

REFERENCES

Bold, Harold C., *Morphology of Plants,* Harper and Bros., New York, 1957.
Fuller, Harry J., and Oswald Tippo, *College Botany,* Revised Edition, Henry Holt and Co., New York, 1955.
Johansen, Donald Alexander, *Plant Microtechnique,* McGraw-Hill, New York, 1940.
Rugh, Robert, *The Frog,* The Blakiston Co., Philadelphia, Pa., 1951.
Wodsedalek, J. E., *General Zoology Laboratory Guide,* Fourth Edition, Wm. C. Brown Co., Dubuque, Iowa, 1955.

PROJECTS

1. Dissect other animals that might be available or easily collected to study reproductive structures. Make a comparative study to see similarities that exist among diverse groups of animals.
2. Dissect out the testis of a grasshopper. Make a smear on a slide. Place a drop of normal saline solution on the smear and observe the sperm.

EXERCISE 16
Mendelian Genetics

MATERIALS

Student Station

dissecting needle
dissecting microscope or magnifying glass
microscope slides and coverglasses
paper towels
compound microscope

Each Group of Four Students

seedling flat with planting soil
package of genetic albino corn seeds
0.9% NaCl solution in dropper bottle
orcein in dropper bottle (add an excess of orcein to 45% acetic acid, then boil and filter)
2 bottles with virgin female *Drosophila* (1 bottle of wild type, 1 bottle of mutant type)
etherizer (A half-pint milk bottle with a cork stopper. Insert a nail wrapped with cotton in the stopper.)

white card
Drosophila salivary gland smear prepared slide

Class Stock

2 bottles of *Drosophila,* blowfly, or housefly larvae (grown in banana agar)
2 bottles of wild type *Drosophila* (grown in cornmeal agar)
2 bottles of mutant *Drosophila* (grown in cornmeal agar). Select mutations on different chromosomes.

OBJECTIVES

1. To grow albino corn, an example of Mendelian genetics.
2. To make a smear of salivary glands to see giant chromosomes.
3. To make a monohybrid and a dihybrid cross with *Drosophila melanogaster,* a useful laboratory organism to illustrate basic genetic principles.

**Part 1
Albino Corn**

Plant corn in the seed flat according to the directions on the package or as directed by the instructor. Set flat aside and observe the seedlings after 1 week. (See Exercise 17.)

**Part 2
Giant Chromosomes of Salivary Glands**

Obtain a large larva from the stock bottle and place it on a slide in a drop of 0.9% salt solution. Pierce the head with a dissecting needle and the posterior end with another needle, then pull slowly in opposite directions. This should release the head and the digestive tract with the salivary glands. With the aid of a dissecting microscope or a magnifying glass, locate the salivary glands (see Fig. 16.1). They are saclike structures on both sides of the digestive tube. When you are sure you have located the salivary glands, carefully cut them away from the surrounding structures. Remove everything from the slide except the salivary glands. Blot the salt solution with paper toweling, and replace the salt solution with a drop of orcein stain. Cover

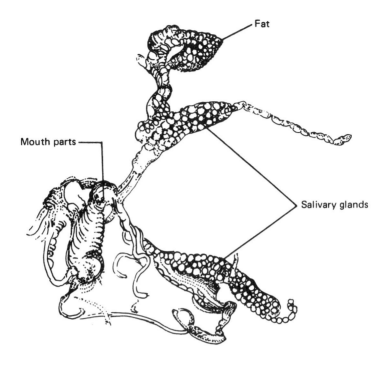

Figure 16.1 Drosophila salivary glands.

the glands with a coverglass. Place a piece of paper toweling over the coverglass and squash the glands. Be careful not to break the coverglass. With the eraser end of a pencil, tap the coverglass a number of times to release the chromosomes from the nuclei. Observe the slide with the low and high powers of the microscope. Compare your slide with a prepared slide of *Drosophila* salivary gland smear.

1. *Describe the appearance of the chromosomes.*

2. *Do you see any dark bands on the chromosomes? What do geneticists think the dark bands are? (Consult a number of reference books for the various theories.)*

3. *Why are these chromosomes larger than chromosomes in other cells?*

**Part 3
Drosophila melanogaster**

Today you will begin two crosses involving the vinegar fly, *Drosophila melanogaster.* The results of these experiments will illustrate basic principles of Mendelian genetics. Work in groups of four.

A. Preliminary Observations

If you are to be successful with these experiments, it is important that you become familiar with the proper handling, anesthetization, and normal and mutant characteristics of the flies, and that you are able to distinguish the sexes.

1. Obtain a stock bottle of wild type (normal) flies.
2. Saturate the cotton pad of an etherizer with ether.
3. Gently tap the stock bottle so that the flies fall to the bottom of the bottle.
4. Remove the plugs from the stock bottle and the etherizer. Quickly invert the stock bottle over the etherizer.
5. Tap the jars a number of times until the flies have dropped into the etherizer.
6. Replace the plugs.
7. Allow the flies to remain in the etherizer only as long as it takes to anesthetize them. Overexposure will result in serious damage to the tissues.
8. Dump the flies onto a white card, and observe them with a dissecting microscope or a hand lens. Move the flies around with the handle end of a dissecting needle or a similar tool. Never use a sharp instrument. Note the red eyes, the straight wings, and the gray body. You will see black bands on the abdomen (the posterior portion of the body). Some flies have a blunt abdomen with three broad bands; these are the male flies. Those that have a pointed abdomen with five narrow bands are the female flies. (See Fig. 16.2.)
9. When you are thoroughly familiar with the morphological characteristics, return the flies to the stock bottle. Insert the plug and invert the bottle so that the flies fall on the plug. It is important that the flies do not remain on the agar before they are revived, since their wings might stick to the agar. Return the bottle to an upright position as soon as the flies revive.

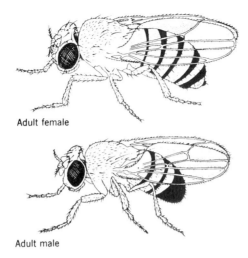

Figure 16.2 Male and female Drosophila.

B. Monohybrid Cross

Each group will be given a jar containing cornmeal-agar and three virgin wild type females.

1. Obtain a stock bottle containing flies with one mutant characteristic, and compare it with the normal fly you observed earlier.
2. Anesthetize the mutant flies and select three males. Observe closely the mutant characteristic and compare it with the normal fly observed earlier.
3. Tap the jar with the normal female flies until they fall to the bottom of the bottle.
4. Remove the plug and quickly drop in the anesthetized male mutant flies.
5. Replace the plug. Remember to invert the bottle until the flies revive.
6. Label the bottle with the type of cross and the date and set aside until the next class period.

C. Dihybrid Cross

Follow the general procedure described for the monohybrid cross. Each group will be given a bottle with three virgin females that exhibit a mutant characteristic, e.g., ebony body. With the females put three males normal for the mutation in the females but exhibiting a mutation for another characteristic, e.g., vestigial wings. The females are normal for the mutation in the males. The genes responsible for these characteristics are on different chromosomes.

1. *Why must virgin females be used in the crosses?*

2. *Why are three pairs of flies used instead of one pair in this experiment?*

SUMMARY

The vinegar fly, *Drosophila melanogaster,* is a useful laboratory organism for studies in genetics. It is small and therefore takes little space in a laboratory. It has a short life cycle (about 15 days) so that many generations may be obtained in a short period of time. The results of experiments with *Drosophila* help to explain basic genetic principles.

REFERENCES

Lawson, Chester A., and Richard E. Paulson, *Laboratory and Field Studies in Biology,* Holt, Rinehart and Winston, New York, 1960.

Royle, Howard A., *Laboratory Exercises in Genetics,* Third Edition, Burgess Publishing Co., Minneapolis, Minn., 1953.

PROJECTS

1. Run additional experimental crosses concurrently with those listed in the exercise. For example, mate a virgin wild type female with a male that has both mutations that were used in the section on dihybrid cross. Compare the results of this F_1 and F_2 with the results obtained from the regular exercise. From your results, determine whether it makes any difference in which sex the mutations are located.
2. Cross a white-eyed male *Drosophila* with a normal virgin female, and cross a normal male with a white-eyed virgin female. Determine whether the results are the same in the F_1 in both crosses.

EXERCISE 17
Human Hereditary Traits

MATERIALS

Class Stock

several booklets of phenylthiocarbamide testing paper (PTC paper)

OBJECTIVES

1. To continue experiments begun last week on corn seedlings and *Drosophila* crosses.
2. To examine and record the incidence of certain human hereditary traits and individual variations.

Part 1
Continuation of Genetics Experiments

A. Corn Seedlings

Examine several flats of corn seedlings and count the numbers of albino and green members. Record the results in Table 17.1.

	Flat 1	Flat 2	Flat 3
Number of green seedlings			
Number of albino seedlings			
Ratio of green to albino			

Table 17.1

1. *The seeds used in this experiment were obtained by crossing parents heterozygous for albinism. What ratio of normal to albino would be expected from such a cross if albinism is recessive to normal?*

2. *How close does this agree with your observed ratios?*

3. *All of the albino seedlings will die soon. Why?*

4. *Why does the trait of albinism continue in some plant populations even though albino members never survive beyond the seedling stage?*

B. Drosophila Genetics

Examine the *Drosophila* crosses that were set up last week. Observe the larvae and pupae. Remove the parents from each vial and dispose of them. Why?

Part 2
The Incidence of Hereditary Traits in Humans

A. Traits

In this exercise you are to record the presence or absence of some of your own hereditary traits in Table 17.2 by circling your probable genotype. These data should then be transferred to a master table for the entire laboratory class as indicated by the instructor. From this master table you can calculate the class percentage for each trait.

Hair Form

Wavy hair results from the interaction between the gene for curly hair and the gene for straight hair. Hence **SS** = straight hair, **Ss** = wavy hair, and **ss** = curly hair. This is an example of incomplete dominance. Record your hair form in Table 17.2.

Widow's Peak

A trait due to a dominant gene **P**, in which the forehead hair line forms a point in the center of the forehead. Straight hair line **pp** is the recessive condition.

Hair Color

Blond or very light hair color **bb** is recessive to dark hair **B−**. Red hair **rr** is recessive to nonred hair **R−**.

Ear Lobes

Ear lobes that do not directly join to the side of the head are termed free ear lobes. This is due to a dominant gene **E**. Attached ear lobes constitute the recessive condition **ee**.

Eye Color

The presence of pigments in the iris causes the eyes to be brown, hazel, green, or other colors and represents the dominant trait **G**. In the recessive condition **gg** the iris appears blue gray.

Freckles

Freckles are dominant **F—** over absence of freckles **ff**.

Tongue Rolling

A dominant gene **T** enables an individual to roll his extended tongue into a V-shape. Inability to do this represents the recessive condition **tt**.

Mid-digital Hair

With the presence of a dominant gene **H**, hair grows on the back side of the middle joints of the fingers. The trait of hairless middle joints is recessive **hh**.

PTC Taster

Individuals who carry the dominant "taster" gene **T** experience a bitter taste when they sample a chemical called phenylthiocarbamide, or PTC for short. Non-tasters are recessive **tt** for this condition. Obtain a piece of PTC test paper, chew it, and note the results.

B. Individual Variation

Select five classmates including a relative if possible and count the number of traits from Table 17.2 that you have in common. Record these data in Table 17.3 and calculate the *percentage* that you have in common with each of them by dividing the number of traits in common by 10.

1. *Did you find 100% agreement in the observed traits with any of your five classmates? If your answer is yes, was the classmate a relative? Why should this matter?*

2. *If you added ten more traits to the list for a total of twenty, would you expect a higher or a lower percentage of traits in common with other people? Explain your reasoning.*

Table 17.2 Hereditary Traits		
Trait	Probable Genotype	Class Percentage
Straight–wavy–curly hair	SS, Ss, ss	
Widow's peak	P____, pp	
Dark hair–blond hair	B____, bb	
Dark hair–red hair	R____, rr	
Free earlobe–attached earlobe	E____, ee	
Nonblue eyes–blue or gray eyes	G____, gg	
Freckles–no freckles	F____, ff	
Tongue rolling–no tongue rolling	T____, tt	
Mid-digital hair–no mid-digital hair	H____, hh	
PTC taster–nontaster	T____, tt	

Table 17.3		
Classmate's Name	Number of Traits in Common	Percent in Common
1.		
2.		
3.		
4.		
5.		

SUMMARY

1. The ratio of normal to albino corn seedlings approximates the 3:1 ratio expected from a simple monohybrid cross in which both parents are heterozygous.
2. It is possible to observe many hereditary traits in humans that conform to simple Mendelian principles of inheritance. A comparison of these traits within a group emphasizes the individuality of human beings.

REFERENCES

Schonberger, Clinton F., *Laboratory Manual of General Biology,* W. B. Saunders Co., Philadelphia, Pa., 1962.

Winchester, A. M., *Biology Laboratory Manual,* Third Edition, Wm. C. Brown Co., Dubuque, Iowa, 1964.

EXERCISE 18
Population Genetics

MATERIALS

Each Group of Four Students

dissecting microscope or hand lens
white card
etherizer
2 bottles containing cornmeal agar
fingerbowl containing mineral oil (for disposing of flies)
bag containing 100 white beans and 100 black beans

Class Stock

ether

OBJECTIVES

1. To continue experiments with *Drosophila*.
2. To study the principles of probability using bean seeds as an example.
3. To calculate gene frequencies and genotype frequencies from class data.

Part 1
Drosophila F_1 Crosses

Examine the bottles containing the crosses started in Exercise 16. Many of the flies should have emerged from the pupae. This week you will examine the F_1 progeny and set up crosses using them for parents.

Etherize the flies and observe them with a dissecting microscope or hand lens. Count the number of flies and write down the morphological traits you are following in these crosses. Include in the data the number of male and female flies present. Make careful observations. Newly emerged flies often appear lightly colored with inconspicuous bands.

Select three females and three males from the bottle set up for the monohybrid cross, and place the flies together in a fresh bottle of corn-agar medium. Dispose of the other flies by dumping them in mineral oil. Record the data in Table 18.1.

Table 18.1

F_1	Number of Wild Type	Number of Mutants	Number of Females	Number of Males
Monohybrid cross				
Dihybrid cross				

Repeat the procedure for the dihybrid cross.

1. *What is the ratio of wild type to mutants in the F_1 monohybrid cross? In the dihybrid cross?*

2. *Suggest reasons for the results you obtained in both crosses.*

3. *Why was it not necessary to use virgin females in the F_1 crosses?*

Part 2
Probability

The probability of an event can be predicted by dividing the number of favorable outcomes for that event by the number of possible times it can occur. Therefore $P = M/N$ where P = probability of an event, M = number of favorable outcomes, and N = number of possible times it can occur. The answer will always be some value between 0 and 1. Because this is a mathematical measure of what can be expected, the values are theoretical. In practice you are not likely to achieve the expected. Large samples are more likely to come closer to the expected than small samples.

A. Probability Concerning Sex in Drosophila

Sex in *Drosophila* involves the X and Y chromosomes. A female has two X chromosomes and a male has an X and a Y chromosome. (See *Fundamental Concepts of Biology.*) Matings in *Drosophila* produce the following results:

	Male Gametes	
	X	Y
Female Gametes X	XX	XY
X	XX	XY

Notice that the progeny are one-half males and one-half females. We can also determine this from the equation $P = M/N$.

$$\text{probability of females} = \frac{2 \text{ (number of females)}}{4 \text{ (number of possibilities)}} = \frac{1}{2}$$

The same is true if we figure the probability for males.

1. *Using the information from Table 18.1, determine how close you came to the expected probability.*

2. *If your results were not close to the expected, suggest a reason.*

168 EXPERIMENTS IN FUNDAMENTAL CONCEPTS OF BIOLOGY

Sex determination is similar in humans and should follow the same probability. In Table 18.2 indicate the number of males and females in your family. How close is it to the expected?

Table 18.2

	Number of Females	Number of Males	Ratio
Single family			
Class			

Obtain data from other members of the class and calculate the ratio of males to females in all the families together. Is the ratio for the class as a whole closer to the expected than for your family?

B. Probability of Single Events

Obtain a bag that contains 100 black beans and 100 white beans. Pick out 8 beans, 1 bean at a time. Be sure to replace the bean each time. Record on Table 18.3 the number of white beans and the number of black beans picked. Calculate the ratio of white to black beans.

Pick out 40 beans, 1 at a time, and record the results. Again be sure to replace the bean each time.

Table 18.3

	White Beans	Black Beans	Ratio
Eight beans			
Forty beans			

1. *What is the expected ratio of black to white beans?*

2. *Which experiment came closer to the expected, 8 beans or 40 beans? Why?*

3. *Why was it important to replace the bean each time before picking the next one?*

C. *Probability of Joint Independent Events*

The probability of independent events happening together can be obtained by multiplying the probability of each independent event. For example, the probability of drawing 2 black beans from the bag at the same time would be ½ · ½ = ¼; 2 white beans, ½ · ½ = ¼; a white and a black, ½ · ½ X 2 = ½. Draw out 40 beans 2 at a time and record the combinations on Table 18.4. Calculate the ratio for each combination. Remember to return the beans to the bag after each draw.

The principles of probability apply to genetics. If there are two genes responsible for a trait, and these two are alleles, there are then three possible combinations of these two genes: **BB, Bb, bb.** In a large population in which there is random mating and mutations are insignificant, we would expect the following to be true:

		Gametes from the Females	
		B (½)	**b** (½)
Gametes from the males	**B** (½)	**BB** (¼)	**Bb** (¼)
	b (½)	**Bb** (¼)	**bb** (¼)

Since the probability of gene **B** occurring in a gamete is ½ and the same is true for **b**, the probability of any combination of the genes occurring together in an offspring is the product of the individual probabilities or **BB** = ½ · ½ = ¼; **bb** = ½ · ½ =

¼; note that **Bb** occurs twice as many times, or ½ · ½ · 2 = ½. These ideas are incorporated into two equations which may be expressed as the Hardy-Weinberg law: Gene frequencies $p + q = 1$; genotype frequencies $p^2 + 2pq + q^2 = 1$. (See

Table 18.4

	Number	Ratio
Black-black		
Black-white		
White-white		

Fundamental Concepts of Biology, Chapter 18.) If the value is known for q^2 (homozygous recessive genotype), the values of the other genotypes and the frequencies of the genes can be obtained by simple mathematical calculations. Study the examples illustrated in the text, Chapter 18, until you understand how to calculate the various values. It should be apparent that if the frequency of a genotype such as **bb** is obtained by multiplying the individual probabilities together, then the frequency of the gene **b** may be obtained by taking the square root of the frequency of **bb**. Since **bb** = ¼, **b** = $\sqrt{¼}$ = ½. In practice, population geneticists obtain the values of gene frequencies and genotype frequencies by application of the Hardy-Weinberg equations and their knowledge of the principles of probability. Part 3 demonstrates the use of these concepts.

Table 18.5

Trait	Class Percentages	Genotypes	Genotype Frequencies	Gene Frequencies
Attached earlobe		ee		e
Free earlobe		Ee		E
		EE		
Blue or gray eyes		gg		g
Nonblue eyes		Gg		G
		GG		
Nontaster		tt		t
PTC taster		Tt		T
		TT		

Part 3
Population Genetics

Using the class percentages in Table 18.2, calculate the genotype and gene frequencies for each of the traits listed on Table 18.5.

1. *In a large population seven out of every ten persons can taste PTC. How close does the class frequency come to the expected frequency?*

2. *If the answer to question 1 is "not close to the expected," can you explain why?*

SUMMARY

1. The use of bean seeds illustrates the principles of probability. The probability of any event is equal to the probability of favorable outcomes divided by the number of possible times it can occur. The probability is expressed as a percentage or ratio.
2. The Hardy-Weinberg equations, $p + q = 1$ for frequencies of genes and $p^2 + 2pq + q^2 = 1$ for frequencies of genotypes, are useful in determining the occurrence of the various genes and genotypes in a population.

REFERENCES

Sinnott, Edmund W., L. O. Dunn, and Th. Dobzhansky, *Principles of Genetics,* Fourth Edition, McGraw-Hill, New York, 1950.

Weisz, Paul B., *Laboratory Manual in the Science of Biology,* Second Edition, McGraw-Hill, New York, 1963.

PROJECTS

1. If you live on a farm or have access to many animals of one species, determine the gene frequencies for various inherited traits that can be observed in the

population. Consult a reference book to compare your results with the expected frequencies.
2. Do the same with a population of plants.
3. Test members of your family and close relatives for ability to taste PTC. Construct a pedigree chart and calculate the frequency of ability to taste PTC in the sample. How does it compare to the expected ratio of 7 out of 10?

EXERCISE 19
Adaptation and Variation

MATERIALS

Class Stock

paper sacks for leaves
rulers
graph paper

OBJECTIVES

1. To continue genetic crosses with *Drosophila*.
2. To observe adaptations among plants and animals found during a brief field trip.
3. To determine the amount of variation in leaf length and the adaptive significance of the variation.

Part 1
Continuation of Drosophila Crosses

Examine your *Drosophila* bottles and remove the F_1 parents. Set the bottles aside for the final observations at the next laboratory period.

174 EXPERIMENTS IN FUNDAMENTAL CONCEPTS OF BIOLOGY

Table 19.1

Adaptation	Probable Function

Plants

Animals

Part 2
Observation of Adaptations in Nature

This portion of the exercise involves a short field trip in the vicinity of the campus. During the field trip observe and record in Table 19.1 the kinds of adaptations and their probable functions. Among plants, watch for growth habits, leaf characteristics, and the general kinds of plants found in different environmental situations. A stream bank or drainage ditch offers a good contrast to a dry field. Among the animals you may encounter, such as insects, note especially the adaptations related to the creature's feeding habits, its locomotion, its possible means of avoiding predators, and other behavioral actions.

Part 3
Variations in Leaf Length

At an appropriate time during your field trip, form a team with three other classmates. Your team is to pick a total of fifty leaves from a tree. Each member should pick randomly from a different side of the tree.

After returning to the lab, measure your leaves in some standard way such as from base to tip. Record your measurements in Table 19.2 in the form of a frequency distribution starting with the shortest leaf length.

Record the data from Table 19.2 on graph paper.

Table 19.2

Leaf Length	Number of Leaves in Each Length Class

176 EXPERIMENTS IN FUNDAMENTAL CONCEPTS OF BIOLOGY

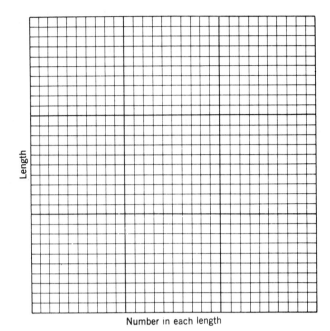

Calculate the mean (average) leaf length for your sample.

$$\text{mean} = \frac{\text{total leaf length}}{\text{total number of leaves}} =$$

1. *Which length class is the most abundant one in your sample?*

2. *How close is this value to the mean which you calculated?*

3. *Do you think that leaf length is an adaptive feature? Why?*

4. *Do you think that the trait of leaf length is inherited?*

5. *In what ways could the environment influence leaf length?*

6. *On any single tree, is the amount of leaf-length variability due to genetic or environmental influence? Explain your answer.*

SUMMARY

1. An alert observer can record a large variety and number of adaptations in plants and animals during a short, local field trip.
2. A sample of leaves taken from a plant shows a range in variation but with one size class more abundant than the others. The variation represents the interaction between heredity and environment. The dominant size class represents the leaf best adapted for existing conditions.

PROJECTS

Prepare leaf-section slides, as directed in Exercise 3, from plants that grow in a dry environment and plants that grow in a wet environment. Compare the slides to see if you can detect structural adaptations within the leaves related to the environment from which they came.

EXERCISE 20
Evolution: Study of Fossils

MATERIALS

Student Station

window-pane glass, 6 in. × 6 in.
cardboard strip, ½ in. × 8 in.
modeling clay

Each Group of Four Students

white cards
dissecting microscope or hand lens
etherizer
fingerbowl containing mineral oil
small leaf

Class Stock

ether
petroleum jelly

180 EXPERIMENTS IN FUNDAMENTAL CONCEPTS OF BIOLOGY

plaster of Paris

fossils for demonstration including samples of the following types: petrifaction, casts, molds

OBJECTIVES

1. To conclude experiments with *Drosophila*.
2. To make an artificial fossil.
3. To examine demonstrations of fossils.

Part 1
Conclusion of Experiments with Drosophila

Examine the bottles containing *Drosophila*. The F_2 flies should have emerged from the pupae by this time. Etherize the flies in the bottle containing the monohybrid cross. Count the number of wild type and the number of mutants. It is not necessary to count the number present of each sex. Enter the results on Table 20.1. Dispose of the flies.

Table 20.1 Monohybrid Cross

Number of Each Phenotype	Phenotypic Ratio

Repeat this procedure for the flies in the bottle containing the dihybrid cross and enter the results in Table 20.2.

Table 20.2 Dihybrid Cross

Number of Each Phenotype				Ratio

1. *What is the expected genotypic ratio for the F_2 flies of the monohybrid cross? Phenotypic ratio?*

2. *Compare the expected phenotypic ratio with the results listed in Table 20.1.*

3. *What is the expected genotypic ratio of the F_2 dihybrid cross? Use a Punnett square to illustrate your answer.*

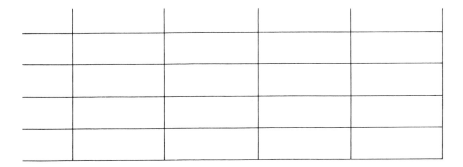

4. *What is the expected phenotypic ratio for the F_2 dihybrid cross? Compare this with the data in Table 20.2*

5. *Account for the appearance of the mutant characteristics in the F_2 even though the F_1 flies were all phenotypically normal.*

6. *The study of the genetics of vinegar flies is of what scientific value?*

Part 2
Artificial Fossil

Spread petroleum jelly over a leaf and place it jelly side up on a piece of glass. Put a strip of cardboard around the leaf, and press clay against the cardboard to hold it in place.

Pour plaster of Paris over the leaf. When the plaster of Paris is dry, remove the leaf. Examine the leaf print with a dissecting microscope or hand lens.

1. *Compare the "fossil" with the leaf. Describe the structural details of the leaf imprinted on the plaster of Paris.*

2. *Describe how a fossil of this type would be produced in nature.*

Part 3
Fossil Types

There are various types of fossils. They are classified according to the process by which they were produced. One process is called *petrifaction;* here the tissues have been replaced by minerals. *Molds* (impressions), another type of fossil, are produced by a hardening of the material surrounding the organism. The organism eventually decays. *Casts* occur if minerals fill the mold to produce a replica of the original organism.

Examine the fossils on display. Enter the name and the type of fossil for each on Table 20.3. Also, include any brief comments you might wish to make.

	Table 20.3			
	Fossil Type			
Fossil Name	Petrifaction	Cast	Mold	Other

1. What type of fossil is the artificial fossil you made in Part 2 of this exercise?

2. Why do you suppose stone tools and weapons of prehistoric man are called indirect fossils?

3. Fossils are direct evidence in support of evolution. What does this statement mean?

4. What is an index or indicator fossil?

SUMMARY

1. The results of the monohybrid and dihybrid crosses conform with the expected Mendelian ratios. Results of genetic experiments with *Drosophila* are significant because they illustrate the basic principles of genetics.
2. The artificial fossil demonstrates one method by which fossils are formed in nature.
3. Petrifaction, molds, and casts are types of fossils, named after the processes that produced them.

REFERENCES

Moore, Raymond C., *Introduction to Historical Geology,* Second Edition, McGraw-Hill, New York, 1958.

Morholt, Evelyn, Paul F. Brandwein, and Alexander Joseph, *Source Book for Science Teaching,* Revised Edition, UNESCO, France, 1962.

Villee, Claude A., *Biology,* Third Edition, W. B. Saunders Co., Philadelphia, Pa., 1957.

PROJECTS

1. Make a number of "fossils" using structures other than leaves, for example, oyster shells, dragonfly wings, floral parts, and stems. Compare the relative ease with which "fossils" can be made from the various structures you use.
2. Plan a field trip to collect fossils. (This should be done only if adequate supervision is available.) Classify the fossils you find according to the fossil types listed on Table 20.3.

EXERCISE 21
Introduction to Ecosystems: A Local Plant Community

MATERIALS

Student Station

notebook
paper or plastic bag for leaves
field clothes
key to local common trees and shrubs

OBJECTIVES

1. To observe some of the structural components of an ecosystem in relation to a local plant community, and to express these observations in the form of an ecological pyramid.
2. To derive and use a simple taxonomic key.

Part 1
Study of a Local Plant Community

This part of Exercise 21 may be done either as a class endeavor or as an independent project as directed by the instructor.

Review Chapter 21 in *Fundamental Concepts of Biology,* which describes structural aspects of an ecosystem, especially abiotic materials, producers, consumers, and decomposers. These are the major features you should attempt to observe on this field trip.

Go to a local plant community for approximately 1 hr to observe and make notes about its structure. Questions similar to those that follow will call your attention to important details. What are the most important abiotic factors in this community? What are the most important producers? What are the major consumers? Is there any evidence of decomposers in the commmunity? Is this a mature community or one that is undergoing succession? The answers to some of these questions will be based on indirect evidence and perhaps even imagination. Nevertheless, you may be surprised at the numerous data you can obtain by alert observing and by knowing what to look for.

While walking about in your community, collect leaves from ten different kinds of trees or shrubs.

After returning to the laboratory, organize your observations into an ecological pyramid as shown below:

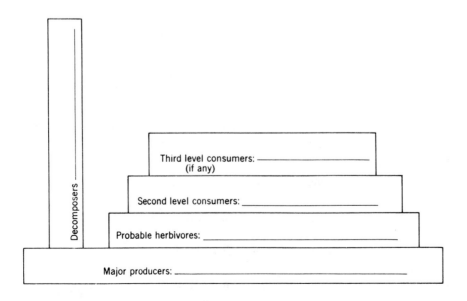

Part 2
Construction and Use of a Simple Taxonomic Key

Anyone working in field biology must learn how to use taxonomic keys to identify different kinds of plants and animals. In addition, the use of a key quickly familiarizes an individual with the major differences and similarities that exist within a group of plants or animals.

Obtain a copy of a key to the common trees and shrubs in your area and use it to "key out" and thus identify your leaves. Notice how the key functions with couplets from which you must choose one or the other alternative at each step throughout the key. Eventually the correct identification is reached. Work with the key until you understand how to use it and can identify most of your leaves.

Take the leaves you collected and divide them into two groups on the basis of a feature that one group lacks and the other group possesses. For example, single-bladed leaves versus lobed leaves, or smooth-margined leaves versus toothed margins. Whatever feature you choose forms the basis for the first couplet in your key. The first couplet directs you to another set of couplets, and this process continues until each leaf is identified. When you are satisfied that your key is workable, let a classmate try it.

Taxonomic keys similar to this have been constructed by specialists for every major group of plants and animals. These keys are extremely helpful to an individual who needs to identify unfamiliar organisms or to an individual who is simply curious to know more about a group of organisms.

SUMMARY

1. A careful and patient observer can estimate the general trophic levels in a plant community. These can be organized into an ecological pyramid.
2. The most accurate and efficient means of identifying plants or animals is usually through the use of a two-branched type of key. The use of this type of key has the additional merits of emphasizing differences and similarities within a group of organisms.

PROJECTS

1. Create a miniature ecosystem in the classroom by setting up an aquarium or a terrarium. In either case it must contain the proper structural features of an ecosystem in order to operate successfully.

    An extension of this idea is to create a sealed system in a gallon bottle. Pond water, some aquatic plants, and a snail or a minnow may persist for a long period of time if sufficient light is provided.

2. Observe several kinds of communities, including a pond if possible, and construct ecological pyramids for each of them.
3. Make a collection of one group of local plants such as ferns, mosses, or shrubs. Identify these and preserve them by pressing between blotters or newspapers until dry. Then mount on stiff white paper. As your collection grows, you can construct a key to the local forms of this particular group.

EXERCISE 22
An Analysis of a Terrestrial Community

MATERIALS

Student Station

dissecting microscope
microscope slides and droppers
fingerbowls

Each Group of Four Students

ball of strong twine
meter stick
notebook
200-ml bottle of preservative
forceps
plastic (or paper) sack
funnel and steelwool

Class Stock

keys for identifying plants and animals

OBJECTIVES

1. To sample a woodland or other natural community.
2. To identify and enumerate some of the major plants and animals in a natural community.

Part 1
Sampling a Terrestrial Community

The class must be divided into small groups in order to conduct the sampling procedure. If a wooded area is to be used, some groups should be designated to count trees, shrubs, and herbaceous plants. The remaining teams will sample square-meter portions of the forest floor for animal life.

A. Trees, Shrubs, and Herbaceous Plants

The technique to be used here will depend partly on the nature of the woodland area. One technique is to lay out a 10 m X 10 m square with stout cord and then count all trees within the square. Each kind of tree or shrub can be given a code letter until it is identified more accurately.

If the woodland is dense, use a smaller sampling area or utilize a transect method. In this method, a student holds two meter-sticks end-to-end and at right angles to his body, like outstretched arms, and walks a predetermined distance (e.g., 20 m) in a straight line through the forest. Any tree touched by the sticks is recorded.

While one or more teams are counting trees, another team (or teams) can set up a 4-square-meter quadrate with heavy twine and count all shrubs in it.

A third team may be directed to count low vegetation in 1 square meter plots.

Trees and shrubs that cannot be identified in the field can often be keyed out in the laboratory from a sample of leaves attached to a small piece of the branch. Herbaceous plants are more difficult to identify and can be designated as grasses, wildflowers, or weeds. Each group, i.e., "trees," "shrubs," and "herbs," is to summarize its findings in a list of plants arranged in order of their abundance.

As arranged in Tables 22.1 to 22.3, the data show the relative abundance of each kind of plant. These figures can be converted into percentages or into numbers per unit area with simple calculations. Consult your instructor for the way in which he wants the data prepared.

B. Animals

Teams of two or three students make satisfactory groups for this type of sampling. Each team marks off a square-meter area on the forest floor. The location of these quadrate sites will be determined by the instructor.

After laying out the quadrate, all visible organisms should be collected and placed in a preservative such as alcohol or formalin. This collection should be made carefully and unhurriedly.

If the sampling area is covered with dead leaves, remove these carefully until you reach the decayed leaf litter and soil area. Look through this material for organisms down to a depth of several inches. A generous sample (200 ml) of this material should be placed in a container for analysis in the laboratory. Normally, two persons will spend 1 to 2 hr collecting the organisms from their sample plot.

Should your quadrate enclose an ant hill or termite nest, collect only a representative sample from each (or move the quadrate to another site!).

Upon completion of the sampling procedure, return your collections to the laboratory and store them away for use at the next laboratory period. Make sure that bottles containing preserved organisms are tightly sealed.

The soil-leaf litter collections must be placed in funnels suspended above a bottle of preservative. A piece of steel wool at the bottom of the funnel prevents debris from falling into the preservative. A light bulb above the funnel gradually dries the contents and causes the organisms to fall into the bottle of preservative. Within a week all of the organisms will be in the preservative.

If plant specimens require identification, they can be held until the next laboratory period by pressing between layers of newspaper. This crude plant press should be stored in a dry location to prevent the specimens from decaying.

Identify the trees and shrubs you collected by using keys provided in the laboratory. Enter your identifications in Tables 22.1, 22.2, 22.3.

The preserved animal specimens should be sorted into fingerbowls and examined. The small forms will require the use of a dissecting microscope. Using keys, identify the animals to Orders. Most of your specimens will be arthropods and a high proportion of these will be insects. List the Orders you find in Table 22.4.

The above information, including the data on plants, provides an assay of the life in the community that you sampled. The techniques used here are not accurate, but they should impress you with the great diversity and abundance of life that exists in a small area.

Table 22.1 Trees per Square Meter Area of Transect Length

Name of Tree   Number

Table 22.2 Shrubs per Four Square Meters

Name of Shrub   Number

Table 22.3 Herbaceous Plants per Square Meter

Name of Plant	Number

Table 22.4

Name of Order	Number Collected	Percentage of Total

1. *If you named the community you sampled by its most abundant* plant *form, what would you call it?*

2. *How would you name it according to the most abundant* animal *form?*

SUMMARY

The plant and animal life in a community can be sampled by relatively simple techniques. Certain forms are more abundant than others. Communities are often named according to several of their most abundant members.

REFERENCES

The *How to Know* series published by Wm. C. Brown Co., Dubuque, Iowa, is helpful in identifying plants and animals.

PROJECTS

1. Compare different kinds of communities according to their floral and faunal members.
2. Make animal quadrate collections at night. Compare the abundance and diversity of nocturnal animals with those collected during the day.

EXERCISE 23
The Pond Community

MATERIALS

Student Station

dissecting microscope
microscope slides and coverglasses
eyedropper
3 small fingerbowls

Each Group of Six Students

small plankton net
4 collecting bottles with lids
plastic pail
fine mesh dipnet
100 ml 10% formalin

Class Stock

Copies of appropriate references for identifying aquatic plants and animals. Several are listed at the end of this exercise.

OBJECTIVES

1. To observe succession stages in a pond community.
2. To collect and identify some of the major forms of animals and plants dwelling in ponds.
3. To relate pond organisms to an ecological pyramid as was done in Exercise 21 for a terrestrial community.

Part 1
Succession Stages in a Pond Community

Small ponds and lakes have short lives—a few months, a few years, a few decades—but eventually they become terrestrial communities. Consequently, a variety of succession stages can be found around and within aquatic environments. As you examine a pond in your area, determine how many of the following stages are present.

1. *Open water and submerged aquatic plants.* Plants in this zone or stage grow under water, often rooted in the bottom. If feasible, collect a plant from this zone and examine it for adaptations related to its underwater habitat.

2. *Zone of plants with floating leaves and flowers.* These are plants like water lilies which are usually attached to the bottom and send long stems or leaf petioles to the surface.

3. *Region of emergent aquatic plants.* This area consists typically of plants like cattails, rushes, and arrowweed. This is a relatively shallow water succession stage in which only the bases of the plants are located underwater. If this stage is absent, floating aquatics from zone 2 may occupy the area.

4. *The marshgrass zone.* Located around the perimeter of many ponds is a band of grasses and sedges called marshgrass.

5. *Shrub zone.* Encrouching on the marshgrass zone are a variety of shrubs such as willow and buttonbush which tolerate wet soil and occasional flooding.

6. *Woodland region.* The shrub zone normally is succeeded by a forest climax in many areas of the country.

In theory the pond should gradually fill with dead vegetation so that each of the plant communities is succeeded by the next one. Eventually the climax stage, usually a woodland stage, is attained.

Observe your pond area and briefly record the appearance of each of the following stages if present.

THE POND COMMUNITY   197

Stage 1:

Stage 2:

Stage 3:

Stage 4:

Stage 5:

Stage 6:

**Part 2
Collection and Identification of Plants and Animals in a Pond Community**

A. Collection of Plants and Animals

Form groups of six. Each group should have the following equipment: plankton net, fine mesh dipnet, 2 pails, and 3 collecting jars with lids.

Three members of the group are to take the dipnet and pail and work around the edge of the pond. Their function is to collect macroscopic pond life such as insect larvae. These are placed in a pail of water for transport back to the laboratory.

The remaining three group members are responsible for collecting plankton. This involves simply passing water through the plankton net until a sufficient number of organisms have been obtained. In shallow water the best collecting technique is to gently dip water into a container and pour it into the plankton net. This should be done carefully to avoid getting silt or mud in the net. If the pond water appears clear and clean, it will be necessary to pour many liters of water through the net to obtain enough organisms for study. If you are working with a scummy pond, only a liter or so is necessary. The plankton collected from the edge of the pond should be placed in one of the collecting bottles with a small amount of water.

To collect plankton from deeper areas of the pond, attach a strong cord to the net so that it can be tossed out into the pond and slowly retrieved. This should be repeated until sufficient amounts of plankton have been collected. Maintain this collection in a separate jar.

The dipnet groups and plankton net groups should exchange duties after making their collections in order to become familiar with both types of collecting.

Upon completing the collections, return all materials to the laboratory and begin sorting and examining the organisms as directed in part B.

B. Identification of Aquatic Organisms

The organisms collected from the pond community are best examined while alive. If this is not practical because of time limitations, preserve your collections in 10% formalin for study during the next laboratory period.

Sort your dipnet collection into fingerbowls. Animals commonly collected are dragonfly and damselfly nymphs, mayfly nymphs, water mites, caddisfly larvae, water beetles, and crayfish. Figure 23.1 illustrates the general body form of some of these organisms. To identify these organisms in greater detail consult a reference book. Most of these forms survive well in an aquarium if you wish to observe them over a period of time.

Pour your plankton net collections into fingerbowls. Examine some of this material with a dissecting microscope and attempt to identify these groups of organisms. Figures 23.2 to 23.6 illustrate a number of common forms, which are described below.

1. *Algae.* You will probably encounter a variety of algae, including single-celled forms, filaments (chains) of cells, colonial forms, and diatoms. The diatoms live in rigid cases or "shells" and frequently are motile. This myriad of microscopic plants constitutes the major producer level in pond communities. See Figs. 23.2 and 23.3.

2. *Protozoans.* These comprise the smallest members visible with the magnification that you are using. The larger motile ones probably consist of ciliates and the smaller ones are likely flagellates. See Fig. 23.4.

3. *Rotifers.* These are frequently abundant in pool water and easily confused with protozoans. There are numerous species and a variety of body forms. Figure 23.5 shows a few of them.

4. *Arthropods: Crustacea.* The following forms, sometimes termed microcrustacea, are common forms of plankton and for the most part are easily recogniz-

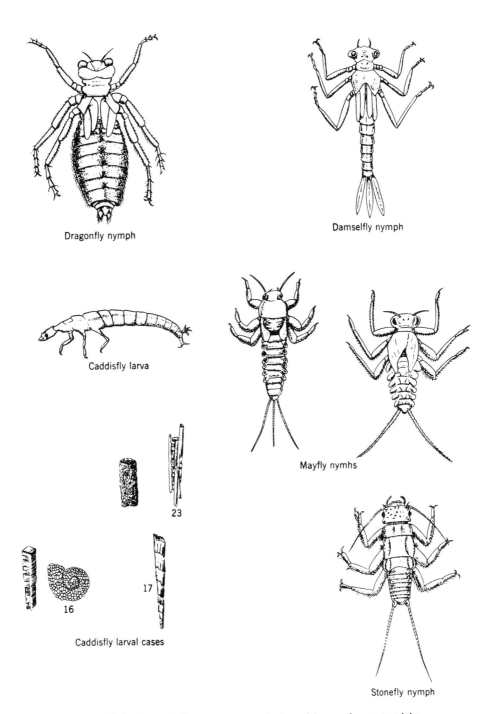

Figure 23.1 Aquatic insects commonly found in pond communities.

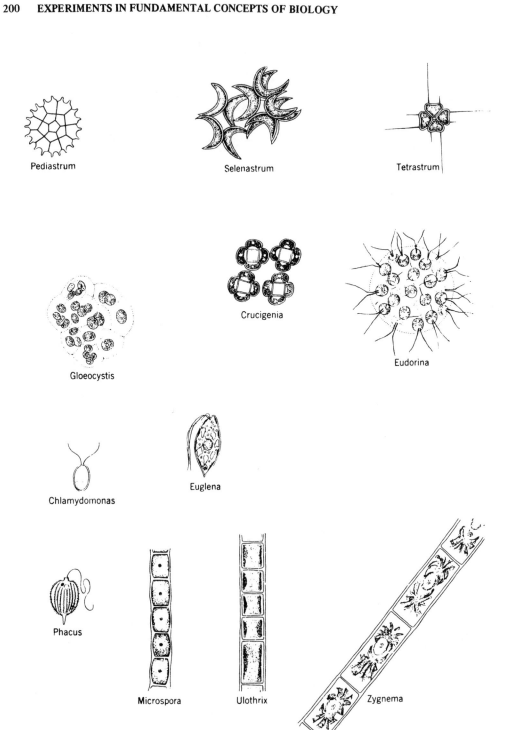

Figure 23.2 Green algae.

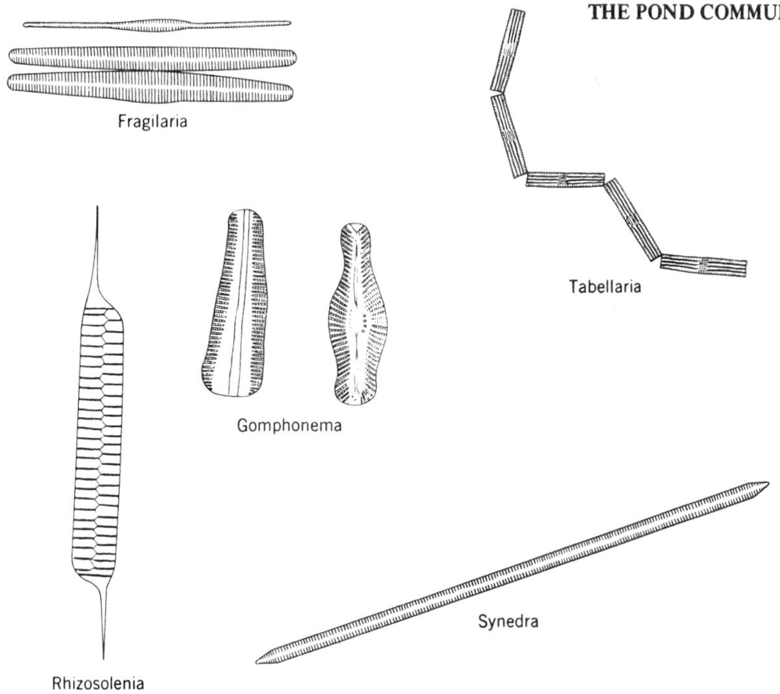

THE POND COMMUNITY 201

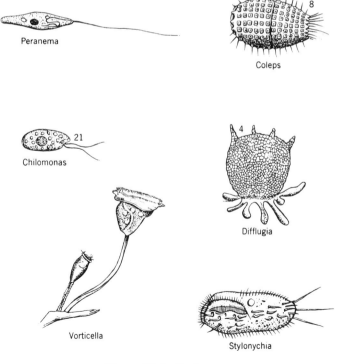

Figure 23.3 Diatoms.

Figure 23.4 Protozoa.

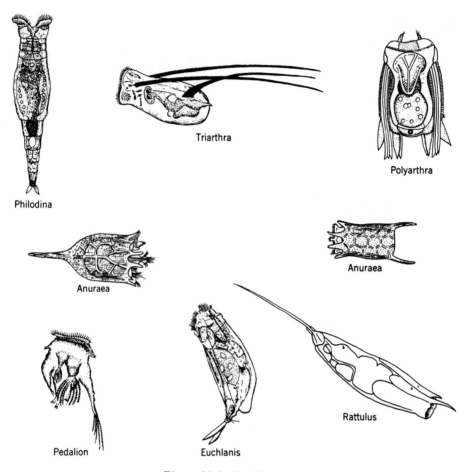

Figure 23.5 Rotifers.

able. Using Fig. 23.6 as a guide, see if you can identify the following forms in your sample:

      a.   Copepoda—copepods
      b.   Ostracoda—ostracods
      c.   Cladocera—water fleas
      d.   Nauplius larvae
      e.   Decapoda—crayfish

    5. *Arthropods: Arachnida.* These include Acarina—water mites—as shown in Fig. 23.7.

THE POND COMMUNITY 203

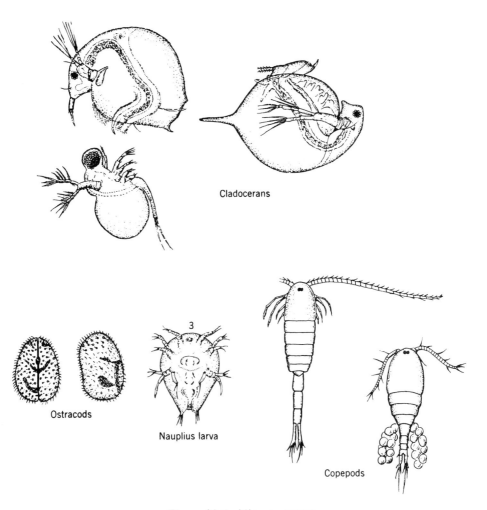

Figure 23.6 Microcrustacea.

Figure 23.7 Water mite.

After you have studied and identified your assemblage of organisms, they can be preserved for future reference in 10% formalin or 70% alcohol.

Part 3
Ecological Pyramid of Pond Organisms

Diagram an ecological pyramid based on the organisms from the pond community.

SUMMARY

1. Various stages of succession can be observed in a pond community. These begin with submerged aquatic plants and end with the climax stage, often a woodland of some type.
2. A large variety of plants and animals inhabit pond communities. Certain groups of these can be recognized with a brief amount of study. These include algae, protozoans, rotifers, copepods, cladocerans, ostracods, nauplius larvae, crayfish, and water mites.
3. Pond life can be organized into an ecological pyramid in which algae form the producer level and protozoa, microcrustacea, and other animals form consumer levels.

REFERENCES

Needham, J. G., and P. R. Needham, *A Guide to the Study of Freshwater Biology,* Fourth Edition, Comstock Publishing Co., Ithaca, N. Y., 1938.
Pennak, R. W., *Fresh-Water Invertebrates of the United States,* The Ronald Press, New York, 1953.
Ward, H. G., and G. C. Whipple, *Fresh-Water Biology,* Reprinted 1945, John Wiley and Sons, New York, 1918; see also Edmondson, W. T., *Ward and Whipple's Fresh-Water Biology,* Second Edition, John Wiley and Sons, New York, 1959.
Welch, P. S., *Limnological Methods,* Blakiston Co., Philadelphia, Pa., 1948.

PROJECTS

1. Perform a quantitative study of the plankton in your pond. To do this, pour a measured quantity of water, such as 10 liters, through a plankton net. In the laboratory place 1 ml of your sample in a plankton counting cell, and count the numbers of different kinds of plankton present. If you find 53 copepods in 1 ml, then you can calculate how many there were in 1 liter of pond water: 53 × 1000/10 = 5300 copepods per liter of pond water in the site where you collected the sample. In this manner you can estimate the density of organisms either collectively or species by species in the pond.
2. Use the preceding technique to follow seasonal changes in plankton populations by sampling at 2- or 3-week intervals for an extended period of time. In such a study you can expect to find considerable fluctuations in the density of each species and in the kinds of species present during different seasons.
3. Determine the physical and chemical factors operating in a pond simultaneously with the studies suggested above. Temperature, light penetration, oxygen content, pH, etc., can all be determined by methods described in Welch (1948) and other works.

Calendar for Preparation of Living Materials

The following calendar lists living materials required in the Exercises and times for planting seeds. Additional materials should be prepared as directed at the beginning of each Exercise.

EXERCISE 1

Plant bean seeds for Exercise 3.

EXERCISE 2

Carrots
Potato
Apples
Liver
Eggs
Salad Oil
Plant bean seeds for Exercise 4

Whole milk
Peanuts, fresh or dry roasted
Yeast
Living *Elodea*

EXERCISE 3

Elodea
Bean seedlings started week of Exercise 1

EXERCISE 4

Leaves: spinach, *Tradescantia,* or privet
Elodea
Plant bean seeds for Exercise 6
Bean seedlings started week of Exercise 2; place in dark three days prior to Exercise 4

EXERCISE 5

Yeast
Frogs or other small animals
Plant radish seeds between moist filter paper for Exercise 6
Plant tomato seeds for Exercise 8

EXERCISE 6

Celery
Frogs
Radish seedlings started week of Exercise 5
Bean plants started week of Exercise 4

EXERCISE 7

Liver

EXERCISE 8

Tomato plants started week of Exercise 5
Two-day-old male chickens
Plant bean seeds for Exercise 10

EXERCISE 9

Plant lima bean seeds for Exercise 11

EXERCISE 10

Bean plants started week of Exercise 8
Prepare *Anolis* lizard cages for Exercise 11
Place 3 male and 3 female *Anolis* lizards in a large terrarium for Exercise 11; each lizard should be coded with dabs of paint for identification
Obtain food for lizards: meal worms, houseflies, or other small insects
Small potted plants for lizard terraria

EXERCISE 11

Anolis lizards, males and females
Soaked corn seeds
Bean seedlings started week of Exercise 9
Onion roots: place onions in water 5 days prior to exercise 12

EXERCISE 12

Onion roots started week of Exercise 11
Irish potatoes
Planaria
Pond water
Plant bean seeds for Exercise 14

EXERCISE 13

Frogs
Mosses
Plant bean seeds for Exercise 14
Germinate bean seeds in wet filter paper 3 days prior to use in Exercise 14

EXERCISE 14

Male frogs
Female frogs (treated with pituitary to induce ovulation)

Pond water known to support frog life
Mature sea urchins (optional)
Bean seeds soaked in water 24 hr
Germinating bean seeds in wet filter paper started week of Exercise 13
Young bean seedlings started week of Exercise 13
Bean seedlings started week of Exercise 12

EXERCISE 15

Flowers
Male and female frogs

EXERCISE 16

Drosophila, blowfly, or housefly larvae
Drosophila for monohybrid and dihybrid crosses
Genetic albino corn seeds

EXERCISE 17

See Exercise in manual

EXERCISE 18

100 white beans and 100 black beans

EXERCISE 19

See Exercise in manual

EXERCISE 20

Small leaves

EXERCISES 21-23

See Exercises in manual

Directions for Pithing a Frog

Hold a frog with one hand so that only the head is exposed. Use your index finger of the same hand to push the head down. Run a finger from your other hand along the middle of the skull until you reach the base of the skull. The opening into the skull is located here. Push a needle into the skull and move the needle from side to side. Remove the needle from the skull and push it into the spinal cord. The needle is in the correct position if the frog's legs stiffen momentarily. Remove the needle. If pithing has been accomplished, the frog will be limp. Repeat procedure if the frog is still stiff.

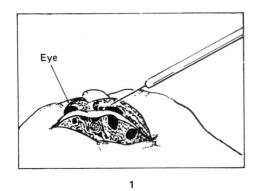

1

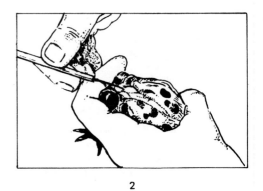

2

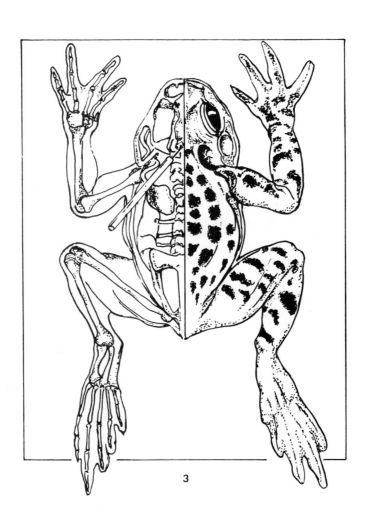

3